NOISY OPTIMIZATION WITH EVOLUTION STRATEGIES

Genetic Algorithms and Evolutionary Computation

Consulting Editor, **David E. Goldberg**
University of Illinois at Urbana-Champaign
deg@uiuc.edu

Additional titles in the series:

Efficient and Accurate Parallel Genetic Algorithms, Erick Cantú-Paz ISBN: 0-7923-7221-2

Estimation of Distribution Algorithms: *A New Tool for Evolutionary Computation*, edited by Pedro Larrañaga, Jose A. Lozano ISBN: 0-7923-7466-5

Evolutionary Optimization in Dynamic Environments, Jürgen Branke ISBN: 0-7923-7631-5

Anticipatory Learning Classifier Systems, Martin V. Butz ISBN: 0-7923-7630-7

Evolutionary Algorithms for Solving Multi-Objective Problems, Carlos A. Coello Coello, David A. Van Veldhuizen, and Gary B. Lamont ISBN: 0-306-46762-3

OmeGA: A Competent Genetic Algorithm for Solving Permutation and Scheduling Problems, Dimitri Knjazew ISBN: 0-7923-7460-6

The Design of Innovation: *Lessons from and for Competent Genetic Algorithms*, David E. Goldberg ISBN: 1-4020-7098-5

Genetic Algorithms and Evolutionary Computation publishes research monographs, edited collections, and graduate-level texts in this rapidly growing field. Primary areas of coverage include the theory, implementation, and application of genetic algorithms (GAs), evolution strategies (ESs), evolutionary programming (EP), learning classifier systems (LCSs) and other variants of genetic and evolutionary computation (GEC). Proposals in related fields such as artificial life, adaptive behavior, artificial immune systems, agent-based systems, neural computing, fuzzy systems, and quantum computing will be considered for publication in this series as long as GEC techniques are part of or inspiration for the system being described. Manuscripts describing GEC applications in all areas of engineering, commerce, the sciences, and the humanities are encouraged. http://www.wkap.nl/prod/s/GENA

GENAGENAGENA
GENAGENAGENA
Genetic Algorithms and
Evolutionary Computation

NOISY OPTIMIZATION WITH EVOLUTION STRATEGIES

by

Dirk V. Arnold
University of Dortmund, Germany

Foreword by

Hans-Georg Beyer

KLUWER ACADEMIC PUBLISHERS
Boston / Dordrecht / London

Distributors for North, Central and South America:
Kluwer Academic Publishers
101 Philip Drive
Assinippi Park
Norwell, Massachusetts 02061 USA
Telephone (781) 871-6600
Fax (781) 681-9045
E-Mail <kluwer@wkap.com>

Distributors for all other countries:
Kluwer Academic Publishers Group
Post Office Box 322
3300 AH Dordrecht, THE NETHERLANDS
Telephone 31 786 576 000
Fax 31 786 576 254
E-Mail <services@wkap.nl>

Electronic Services <http://www.wkap.nl>

Library of Congress Cataloging-in-Publication Data

Arnold, Dirk V., 1970-
Noisy optimization with evolution strategies / Dirk V. Arnold.
p. cm. -- (Genetic algorithms and evolutionary computation)
ISBN 1-4020-7105-1 (alk. paper)
1. Genetic algorithms. 2. Mathematical optimization. 3. Noise. I. Title. II. Series.

QA402.5 .A76 2002
519.7--dc21 2002067470

Printed on acid-free paper.
Printed in the United States of America

Contents

Foreword

During the past decade evolutionary computation has reached a status of broad acceptance in all branches of engineering and the sciences. A main reason for its big success — especially in the engineering disciplines and in computer science — lies in its ease of use: The underlying variation-selection principle is fairly simple, it does not necessarily require problem-specific knowledge, and it is implemented easily. On the other hand, the results are often surprisingly good and quite competitive compared with problem-specific optimization methods (if available or known at all) — at least at first glance.

Being successful in solving specific optimization problems, the average user does not usually ask for the deeper reasons of their (fast) success. Thus, evolutionary computation evolved itself and appears by now as a mainly application-driven discipline having certain deficits in its theoretical basis. Of course, analyzing evolutionary algorithms is far more difficult than analyzing problem-specific optimization algorithms, which are often theory-driven and designed in such a way that they are amenable to a mathematical analysis. (Actually, this is the main difference between evolutionary computation and mathematical optimization.) Therefore, contributions to a theory of evolutionary computation which allow for an assessment of the algorithms' performance are highly desirable and of importance for being accepted as part of the "exact" sciences. If such contributions also have implications on practice, then we are in an ideal situation.

Dirk Arnold's book belongs to this category. It addresses one of the most pressing and interesting topics in evolutionary computation research — the performance of evolutionary algorithms in uncertain environments, i.e., applications where the objective functions are disturbed by noise. Noise is a common phenomenon in many applications, even though it is mostly neglected or removed already at the stage of defining the objective function. When applying evolutionary algorithms, however, there is no need to remove the noise explicitly.

Considering real-valued evolution strategies (ES), Dr. Arnold shows in this interesting book that ES can deal with noise efficiently. That is, his theoretical results indicate that using a recombining population is more efficient than decreasing the noise by averaging objective function values (the usual procedure used in the optimization community). Furthermore, his investigations reveal a superiority of population-based recombinative general purpose ES over state-of-the-art direct optimization algorithms like simplex methods, response surface techniques, and implicit filtering which were explicitly designed for dealing with noise.

Obviously, Dr. Arnold has found a field where ES excel — the optimization under uncertainties. Apart from this interesting message with its practical implications, this book also contributes to the theory of order statistics. It will be useful for both theoreticians as well as practitioners in the field of optimization and statistics. It also bears the potential of bridging the still existing gap between evolutionary optimization and the mathematical optimization community. I wish this book a wide dissemination.

Hans-Georg Beyer
Bochum, Spring 2002

Acknowledgments

This book summarizes the results of three years of work on evolution strategies for optimization in continuous, noisy search spaces. During this period of time, I was a member of Hans-Georg Beyer's research group on theoretical aspects of evolutionary computation at the Chair of Systems Analysis at the University of Dortmund. With his work on evolution strategies, Hans-Georg not only paved the way for this book, but he also acted as an advisor and took active part in and contributed to the research in numerous ways. I would also like to thank the head of the Chair of Systems Analysis, Hans-Paul Schwefel, for continued support and for his friendly interest in my work. Finally, financial support of the author by the Deutsche Forschungsgemeinschaft (DFG) under grants Be1578/6-1 and Be1578/6-3 during the time that the book was written is gratefully acknowledged.

Chapter 1

INTRODUCTION

Optimization is a branch of the computational sciences that is concerned with determining optimal solutions to certain mathematical problems. Frequently, such mathematical problems arise as models of physical reality in all areas of the natural sciences, in engineering, economics, and management. Solving an optimization problem is to find a combination of parameter values which optimizes a given quantity, possibly subject to restrictions on the allowed parameter ranges. The quantity to be optimized is known as the *objective function*, restrictions on the allowed parameter ranges are commonly referred to as *constraints*.

For many optimization problems, finding a global optimizer, i.e. a combination of parameter values that satisfies all constraints and that has the property that no better combination of parameter values that satisfies all constraints exists, is effectively impossible. Frequently, the best one can settle for are local search strategies that attempt to iteratively improve the quality of combinations of parameter values that we refer to as *candidate solutions*. The quality of a candidate solution is determined directly by the objective function of the problem at hand. Local search strategies use information on the objective function only from within a neighborhood of the current location in parameter space of the search. The objective is not to generate a sequence of candidate solutions that provably converges to a global optimizer, but rather to improve the quality of the candidate solutions as rapidly and with as little effort as possible. The effort involved in a search is usually quantified by the number of objective function evaluations performed. Local search strategies can be evaluated quantitatively and can be compared on specific objective functions by considering rates of improvement of the objective function values over time.

Noise is a common factor in most real-world optimization problems. In particular, the evaluation of the objective function value of a candidate solution may be subject to disturbances. We refer to an objective function in combination with a noise model as a *fitness environment*. Sources of noise include, to name but a few, physical measurement limitations, stochastic simulation models, incomplete sampling of large spaces, and human-computer interaction. While it may improve global convergence properties of some algorithms in multimodal settings, noise is usually detrimental to the local performance of search strategies. It can lead to the replacement of candidate solutions with inferior ones, or to the rejection of superior candidate solutions based on comparisons of noisy objective function values. Excessive amounts of noise can result in divergence of search strategies even on the most simple objective functions.

Evolutionary algorithms are general, nature-inspired heuristics for numerical search and optimization. The three major variants distinguished by Bäck [12] — genetic algorithms, evolutionary programming, and evolution strategies — have originated independently and differ in their particulars, but share the same basic paradigm. Starting from an initial set of candidate solutions, in an iterative process, new candidate solutions are generated from existing ones by means of variation, and selection serves to drive the set of candidate solutions towards increasingly better regions of the search space. Variation is achieved by means of recombination — the act of combining several candidate solutions to form a new one — and mutation — the random modification of parameter values. Adopting the usual terminology, we refer to the objective function as the *fitness function*, to time steps as *generations*, to the set of candidate solutions as a *population*, and to existing and newly generated candidate solutions as *parents* and *offspring*, respectively. Strategies with populations of size one are called *single-parent strategies*, and strategies with populations of size greater than one are referred to as *multi-parent strategies*. When we say that a candidate solution has a higher fitness than another, we mean that it has a higher objective function value if the task at hand is maximization, and that it has a lower objective function value if the task is minimization.

Industrial applications of evolutionary algorithms date back at least to the 1960s, and areas of application today include management, control, design, scheduling, pattern recognition, and decision making. A host of international conferences and several international journals are devoted to the field of evolutionary computation. In many instances, evolutionary algorithms have proven to be robust and are frequently employed to solve challenging problems where traditional methods are prone to failure, such as optimization problems with highly discontinuous objective functions or where only unreliable data is available. Major reasons for the widespread use of evolutionary algorithms are their universal applicability and the relative ease with which the underlying paradigm is understood and implemented.

The subject of investigation of the present book is the local performance of evolution strategies in continuous search spaces with fitness measurements corrupted by noise. The issues that arise in continuous search spaces are sometimes quite different from those encountered in discrete search spaces. Differential calculus and the theory of order statistics provide powerful mathematical tools, and the work by Rechenberg [66, 67], Schwefel [76, 77] and Beyer [27] has established a framework within which the analyses can proceed. The goal of such research is to determine how the performance of evolution strategies scales with parameters of the problem — such as the dimensionality of the search space or the noise strength — and of the search strategy — such as the population size or the mutation strength. Scaling laws allow for a comparison of different variants of the strategies, provide guidelines for tuning evolution strategies for maximum performance, and offer insights and an understanding of the behavior of the strategies that goes beyond what can be learned from mere experimentation.

An evolution strategy applied to a search problem forms an iterated, stochastic, nonlinear, dynamical system. To study the dynamics, specific fitness environments need to be considered. Except for very simple search problems, the analysis is prohibitively difficult. Simple fitness environments can serve to highlight differences between strategy variants, can help develop intuition with respect to the working principles of the operators the strategy employs, and can motivate the invention of new tools for the analysis of the algorithms that can then be used in more complex cases. If the purpose of the analysis is to study the influence of parameters such as the search space dimensionality, the population size, or the noise strength, most can be learned by considering the most simple nontrivial fitness environments as they lead to the most transparent results.

Focus in the present book — as in much of the research on evolution strategies in the past — is on unconstrained, real-valued search spaces with objective functions $f : I\!R^N \to I\!R$, and we assume a Gaussian noise model. In particular, we will use a linear function with constant noise strength to study distributed populations, and we will employ a spherically symmetric objective function with fitness-proportionate noise strength in most of the other parts of this book. Both fitness environments have in common that due to their symmetries, after some time, the evolution strategy reaches a state that is described by time-invariant probability distributions. Initialization effects and termination criteria are without relevance and can be disregarded. Especially the spherically symmetric function — albeit mostly in the absence of noise — has been used extensively in the past and plays a role in the theory of evolution strategies that is comparable in many ways to the role that the counting ones function plays for genetic algorithms in discrete search spaces.

Previous to the work presented here, a number of publications had dealt with the performance of evolution strategies in the presence of noise. In a very early work, Rechenberg [66] determined the progress rate of a $(1+1)$-ES — a simple single-parent evolution strategy that generates a single offspring candidate solution per time step — on a noisy linear function with constraints that he referred to as a corridor. Beyer [18] was able to determine the progress rate of the $(1+1)$-ES on the noisy sphere. His analysis presumed that the fitness of a surviving candidate solution is reevaluated in every time step. In that same reference as well as in a book by Rechenberg [67], an analysis of the behavior of the $(1, \lambda)$-ES — a single-parent strategy that generates several offspring candidate solutions in every time step — on the noisy sphere was presented. Rescaled mutations as a means of improving the performance of the $(1, \lambda)$-ES in the presence of noise were studied by Rechenberg [67] and by Beyer [24, 26]. Bäck and Hammel [13, 41] empirically investigated the performance of evolution strategies in the presence of noise. Their experiments involved multi-parent strategies and mutation strength adaptation mechanisms, but did not provide explanations for the behaviors that could be observed. Rechenberg [67] also conducted computer experiments with multi-parent evolution strategies in the presence of noise without being able to fully interpret their results. Analytical results for multi-parent strategies were restricted to noise-free environments. In particular, Beyer [20, 21] studied the performance of the (μ, λ)-ES and the $(\mu/\mu, \lambda)$-ES — multi-parent strategies without recombination and with both global discrete and with global intermediate recombination — on the sphere. A result for global intermediate recombination on the sphere was also given by Rechenberg [67]. Analytical results with regard to mutation strength adaptation mechanisms were limited to an investigation by Beyer [22] of the performance of mutative self-adaptation of the $(1, \lambda)$-ES on the sphere and a study of mutative self-adaptation under random selection by Beyer and Deb [30]. Investigations of the performance of other types of evolutionary algorithms, such as studies by Fitzpatrick and Grefenstette [37], Rattray and Shapiro [65], and Miller and Goldberg [57], concentrate on discrete search spaces and therefore differ both in their focus and in the methods employed. It is interesting to note, however, that in spite of the different methodological approach, Beyer [26] found that some of the equations derived in the work by Miller and Goldberg are quite similar with respect to their functional structure to the results derived in continuous search spaces. It can be hoped that the links that seem to exist between continuous and discrete optimization will be made more explicit in the future.

The present text subsumes many of the results on evolution strategies in continuous, noisy search spaces obtained previously as well as it extends them in a number of ways. In Chapter 2, we first introduce the $(\mu/\rho \overset{+}{,} \lambda)$-ES with isotropic normal mutations and briefly discuss various mechanisms for the adaptation of the mutation strength. With the line and the sphere we then introduce two fitness environments that will be used throughout this book. Subsequently, we turn to a brief discussion of local performance measures. Section 5 is of particular importance as it introduces a simple model for the spherical environment that forms the basis for most of the calculations in the following chapters. By virtue of making the process of modeling the behavior of evolution strategies on the sphere for high search space dimensionality explicit, we have achieved a conciseness in the presentation that we hope contributes to the accessibility of the argument.

Chapter 3 summarizes results from the analysis of the local performance of the $(1 + 1)$-ES in the spherical environment presented in [10]. The analysis differs from a previous study by Beyer [18] in that we do not assume that the fitness of surviving candidate solutions is reevaluated. The resulting overvaluation of the parental fitness is identified as a factor that strongly influences the performance of the strategy and that has largely been overlooked in the past. Its effects are studied in detail and implications for the problem of mutation strength adaptation are discussed.

In Chapter 4, the local performance of the (μ, λ)-ES in the linear environment is studied. Preliminary results have been published in [9]. In contrast to the other strategies analyzed in this book, a population that is distributed in search space needs to be modeled, thereby greatly increasing the mathematical difficulties involved. At the same time, the results are of great interest as they can serve to explain the benefits of distributed populations of candidate solutions for optimization in the presence of noise. Implications for optimization on the sphere are discussed as well.

Chapter 5 starts with an analysis of the performance of the $(\mu/\mu, \lambda)$-ES with intermediate recombination in the spherical environment. The results of this analysis, some of which have been published in [7, 8], neatly generalize a number of previously known results and reveal how genetic repair resulting from recombination is especially beneficial in the presence of noise. However, while the results are strikingly simple, their accuracy is satisfactory only in very high-dimensional search spaces and for small population sizes. Therefore, in Section 2, we abandon the simple model of the spherical environment introduced in Section 5 of Chapter 2 and greatly improve the accuracy of our predictions by including terms in the analysis that had been neglected before. The approach taken is similar to that pursued in [11]. Interesting conclusions with regard to optimal population sizes are drawn. Then, in Section 3, we study the effects of noise on the cumulative mutation strength adaptation mechanism by Hansen and Ostermeier [42, 44]. Based on the simple model from Section 5 of Chapter 2, simple results that quite accurately reflect the qualitative behavior of the strategy are obtained. By incorporating the terms that had already led to the improved results in Section 2, good quantitative agreement with measurements is achieved and again conclusions with regard to optimal population sizes can be drawn.

Chapter 6 presents an empirical comparison of the performance of a number of search strategies on the noisy sphere. In addition to a $(\mu/\mu, \lambda)$-ES with cumulative mutation strength adaptation, the direct pattern search algorithm of Hooke and Jeeves [47], the simplex algorithm of Nelder and Mead [58], the multi-directional search method of Torczon [86], the implicit filtering algorithm of Gilmore and Kelley [38], a variant of the implicit filtering algorithm that makes use of the simultaneous perturbation gradient approximation of Spall [78], and the evolutionary gradient search method of Salomon [74] are considered. Interesting differences in the scaling behaviors of the strategies are observed and strengths and weaknesses of the respective strategies are identified.

Chapter 7 concludes with a summary and discussion of the main results and with goals for future research.

Mathematical details have been omitted from many of the calculations and have been deferred to the appendix so as to increase the readability of the text and to minimize the technical difficulties for the reader mainly interested in results. Appendix A is a very brief recapitulation of some statistical basics that we recommend as a reminder of some basic concepts as well as a first reference. Appendix B lists and proves a number of identities that are useful in many of the calculations throughout this book. Appendix C provides detailed calculations that have been omitted from Chapter 3 and that are less important for the overall understanding of the text. Finally, Appendix D derives some results required in Chapters 4 and 5 as well as it introduces noisy order statistics as an important tool for the analysis of multi-parent strategies in the presence of noise.

Chapter 2

PRELIMINARIES

In this chapter, we define the strategies and fitness environments to be analyzed in the remainder of this work. It serves the purpose of introducing basic definitions and terminology, and of preparing the reader for the analyses presented in the following chapters. We do not attempt to introduce evolution strategies in their full generality, but rather restrict ourselves to real-valued search spaces with fitness functions $f : I\!R^N \to I\!R$ and to $(\mu/\rho \overset{+}{,} \lambda)$-ES with intermediate recombination and with isotropic normal mutations. This choice of strategy is motivated both by that is relatively amenable to mathematical analysis and by its pervasiveness in the literature. References to other strategy variants are included where appropriate.

In Section 1, we introduce the $(\mu/\rho \overset{+}{,} \lambda)$-ES in its most basic form. Mutation strength adaptation mechanisms as important integral components of evolution strategies are discussed in Section 2. In Section 3, two simple fitness environments that occupy a prominent place in the literature — the linear environment and the spherical environment — are introduced and motivated, and in Section 4, measures for the local performance of evolution strategies are discussed. Section 5 introduces a simple model for the spherical environment.

1. The Basic $(\mu/\rho \overset{+}{,} \lambda)$-ES

Evolutionary algorithms strive to drive populations of candidate solutions to an optimization problem towards increasingly better regions of the search space by means of variation and selection. A $(\mu/\rho \overset{+}{,} \lambda)$-ES operates with a population $\mathcal{P}$ of μ candidate solutions. Time proceeds in discrete steps and is indicated by a superscript $^{(t)}$ where necessary. In every time step t, a set $\mathcal{Q}^{(t)}$ of λ candidate solutions is created from $\mathcal{P}^{(t)}$ by means of the variational operators of recombination and mutation. The symbol ρ indicates the number of parental candidate solutions involved in the creation of each single offspring

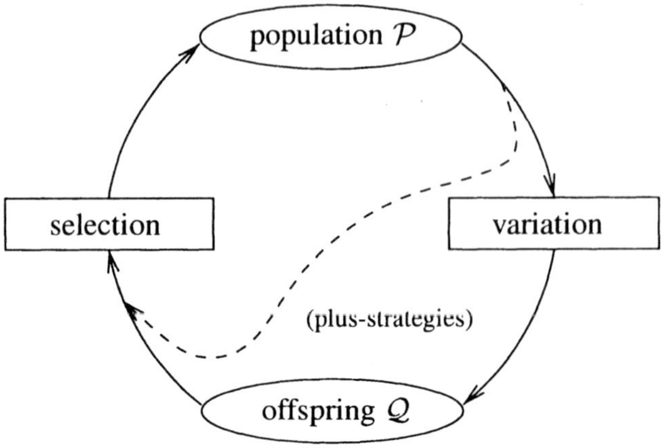

Figure 2.1. The basic evolution loop. Recombination and mutation operators are applied to a population $\mathcal{P}$ of candidate solutions to generate a set $\mathcal{Q}$ of new candidate solutions. Selection is then used to reduce the population to its original size. For comma-strategies, selection is from the set of offspring $\mathcal{Q}$, for plus-strategies it is from the union $\mathcal{P} \cup \mathcal{Q}$ as indicated by the dashed line.

candidate solution. The candidate solutions to form the population $\mathcal{P}^{(t+1)}$ of time step $t+1$ are selected on the basis of their individual fitness — depending on the selection type — either from $\mathcal{P}^{(t)} \cup \mathcal{Q}^{(t)}$ or from $\mathcal{Q}^{(t)}$. Figure 2.1 illustrates the basic procedure. We do not bother writing down initialization schemes and termination criteria as they are irrelevant for the analyses presented in what follows. The following paragraphs describe the operators used for variation and selection in greater detail.

Variation is crucial for preventing stagnation of the evolutionary search. It can be considered a source of innovation and is usually undirected. The process of creating an offspring candidate solution involves *recombination* and *mutation*. Recombination is a process in which $\rho \leq \mu$ parental candidate solutions are selected at random and their centroid is computed. Mutation consists in adding a random vector drawn from an isotropic normal distribution to that centroid. For a population $\mathcal{P} = \{\mathbf{x}_1, \ldots, \mathbf{x}_\mu\}$, the set $\mathcal{Q}$ thus consists of offspring candidate solutions

$$\mathbf{y}_i = \frac{1}{\rho} \sum_{j=1}^{\rho} \mathbf{x}_{i_j} + \sigma \mathbf{z}_i \qquad i = 1, \ldots, \lambda, \tag{2.1}$$

where the indices i_j are independently drawn with equal probability from $\{1, \ldots, \mu\}$, and where the *mutation vectors* $\mathbf{z}_i$ consist of N independent components drawn from a standardized normal distribution. The scalar variable σ determines the expected distance of an offspring candidate solution from the centroid of its parents and is referred to as the *mutation strength.* Throughout the present work, we are concerned only with the two special cases $\rho = \mu$ and $\rho = 1$. In the former case, we stipulate that the indices i_j be drawn from $\{1, \ldots, \mu\}$ without replacement. As a consequence, all μ parents are involved in the creation of every single offspring candidate solution and recombination is in fact deterministic. We refer to this case as *global intermediate recombination.* As all mutations are applied to the centroid of the parental population the analysis is simplified significantly. In the latter case, we adhere to common practice and stipulate that the indices i_j be drawn from $\{1, \ldots, \mu\}$ with replacement. Recombination then simply consists in randomly choosing a parent. We follow common convention and refer to the $(\mu/1 \overset{+}{,} \lambda)$-ES as $(\mu \overset{+}{,} \lambda)$-ES and sometimes call it a strategy without recombination.

Selection is the goal-directed component of the evolutionary search and requires the evaluation of the fitness of the candidate solutions. It is deterministic, with the $(\overset{+}{,})$ symbolism denoting the two mutually exclusive types of selection. Using *plus-selection*, the μ best of the $\mu + \lambda$ candidate solutions in $\mathcal{P}^{(t)} \cup \mathcal{Q}^{(t)}$ are selected to form $\mathcal{P}^{(t+1)}$. Using *comma-selection*, the life span of a candidate solution is restricted to a single time step and it is the μ best of the λ candidate solutions in $\mathcal{Q}^{(t)}$ that form $\mathcal{P}^{(t+1)}$. Obviously, comma-strategies require $\lambda \geq \mu$. We refer to the fraction μ/λ as the *truncation ratio.* With the single exception of the $(1+1)$-ES, all strategies considered in this book are comma-strategies.

For comma-strategies, given a set $\{\mathbf{y}_1, \ldots, \mathbf{y}_\lambda\}$ of offspring candidate solutions, irrespective of whether the task at hand is minimization or maximization, we denote the kth best candidate solution (i.e. that candidate solution with the kth largest fitness value if the task is maximization and that candidate solution with the kth smallest fitness value if the task is minimization) by $\mathbf{y}_{k;\lambda}$. In Section 3 we will introduce the notion of fitness noise and distinguish between the ideal fitness of a candidate solution and its measured fitness. In the presence of noise, $\mathbf{y}_{k;\lambda}$ denotes the offspring candidate solution with the kth best measured fitness value. Selection causes the population of the succeeding time step to be $\{\mathbf{y}_{1;\lambda}, \ldots, \mathbf{y}_{\mu;\lambda}\}$.

Clearly, the variation and selection operators thus defined are but a small subset of the great number of variants that have been suggested and that are being used. The choice of operators outlined above is motivated both by that they are in widespread use and fairly standard and by their relative mathematical tractability. Among the more common alternatives is *dominant* or *discrete recombination.* Beyer [21] introduced *surrogate mutations* as a tool for modeling discrete recombination. Several extensions, such as Cauchy distributed mutations [73], spatially distributed populations [70], or co-evolutionary selection schemes [46] aim at improving global search properties of evolutionary algorithms. Pointers to a great number of such variants can be found in the work by Bäck, Hammel, and Schwefel [14].

Perhaps the greatest practical shortcoming of the strategy outlined above is its reliance on isotropic mutations. Most fitness environments will not exhibit similar scales in different dimensions and will be non-separable. Often, correlated mutations will be necessary to achieve satisfactory performance. In practical applications, the scalar variable σ in Equation (2.1) is therefore to be replaced by an $N \times N$ matrix $\mathbf{S}^{\mathrm{T}}$. According to Rudolph [71], mutation vectors are then normally distributed with positive definite covariance matrix $\mathbf{S}^{\mathrm{T}}\mathbf{S}$. The matrix $\mathbf{S}^{\mathrm{T}}$ can better be adapted to the local structure of the fitness landscape. In the present work, we do not consider correlated mutations as they render the strategies too complicated for mathematical analysis. However, we also note that the restriction to isotropic mutations may not be as severe as it seems. Ideally, a mutation strength adaptation scheme is able to adapt the matrix $\mathbf{S}^{\mathrm{T}}$ such that locally, fitness functions are rescaled into the spherical function to be introduced in Section 3. According to Hansen and Ostermeier [44], using their cumulative mutation strength adaptation mechanism, "Any convex-quadratic function is rescaled into the sphere function". Analyses of the performance of evolution strategies on the sphere can therefore be expected to bear relevance to other fitness functions as well.

In preparation of the following sections, let us define the *population centroid* of a population of candidate solutions $\mathbf{x}_i$, $i = 1, \ldots, \mu$, as

$$\langle \mathbf{x} \rangle = \frac{1}{\mu} \sum_{i=1}^{\mu} \mathbf{x}_i,$$

and let the *population fitness* be the fitness $f(\langle \mathbf{x} \rangle)$ of the population centroid. Furthermore, the *progress vector* $\langle \mathbf{z} \rangle$ is defined by the equation

$$\langle \mathbf{x} \rangle^{(t+1)} = \langle \mathbf{x} \rangle^{(t)} + \sigma^{(t)} \langle \mathbf{z} \rangle^{(t)}. \tag{2.2}$$

Thus, for example, for a $(\mu/\mu, \lambda)$-ES the progress vector is the average of those mutation vectors that generate the offspring candidate solutions included in the next time step's population. For the $(1+1)$-ES the progress vector equals the mutation vector if the offspring is accepted and it equals the zero vector if the parent is retained.

2. Mutation Strength Adaptation

Much of the present work is concerned with evaluating the local performance of evolution strategies either for fixed mutation strength or under the assumption of a certain dependence of the mutation strength on the location in search space.

Practically, the mutation strength σ (or, in case of non-isotropic mutations, the matrix $\mathbf{S}^{\mathrm{T}}$) needs to be adapted in the course of the evolutionary search. An ill-adjusted mutation strength can slow down progress by orders of magnitude if the mutation strength is too low, or lead to divergence if the mutation strength is too high. A mutation strength adaptation component is therefore an important integral part of evolution strategies.

Presumably the first mutation strength adaptation scheme was proposed for the $(1+1)$-ES in the seminal book by Rechenberg [66] . For the $(1+1)$-ES, the *success probability* is the probability that an offspring candidate solution is superior to its parent. Rechenberg's scheme relies on the observation that for the fitness functions he investigated, the success probabilities in case of optimally adjusted mutation strength are in a range of values centered at about one fifth, and that generally increasing the mutation strength reduces the success probability and vice versa. Thus, Rechenberg's recommendation was to monitor success probabilities by averaging over a number of time steps, and to increase the mutation strength if the observed estimate of the success probability exceeds 0.2 and to decrease the mutation strength if the success probability is below 0.2. Schwefel [77], p.112, suggests the following implementation of that rule:

> After every n mutations, check how many successes have occurred over the preceding $10n$ mutations. If this number is less than $2n$, multiply the step lengths by the factor 0.85; divide them by 0.85 if more than $2n$ successes occurred.

We will discuss the applicability of this rule in noisy fitness environments in Section 3.3 of Chapter 3.

For multi-parent strategies, at least three different approaches to adapting mutation strengths have been proposed:

Nested ES, propagated by Herdy [45] and Rechenberg [67], adjust strategy parameters on a meta level by means of evolutionary optimization. Several populations, each one with their own strategy parameter settings, compete with each other for survival. After a number of time steps, the respective progress of the different strategies is determined. The strategy parameter settings of those populations that have achieved the highest progress are used as a basis for generating strategy parameter settings for the next round of competition by means of recombination and mutation. Clearly, nested ES lend themselves well to parallel implementation.

On the downside, there are strategy parameters on the meta level, such as the length of isolation periods and the number of strategy parameter settings to be evaluated, that need to be set appropriately. Currently, there is little knowledge regarding the influence of the meta level strategy parameters on optimization performance. In addition, efficiency concerns have not been addressed. To our knowledge, a mathematical analysis has not yet been attempted.

Mutative self-adaptation, due to Rechenberg [66, 67] and Schwefel [76, 77], includes the strategy parameters into the optimization process at the same hierarchical level as the object parameters of the problem. Different candidate solutions have differing strategy parameter settings. Assuming that favorable strategy parameter settings are more likely to generate successful offspring than unfavorable ones, selection of favorable strategy parameter settings is then a by-product of evolution.

In practice, however, mutative self-adaptation is not without problems. While Beyer [22] has shown that it can lead to a near optimal mutation strength for the $(1, \lambda)$-ES on the sphere, for the $(\mu/\mu, \lambda)$-ES with $\mu > 1$ mutative self-adaptation usually leads to mutation strengths that deviate substantially from their optimal values and therefore to suboptimal performance. Mutative self-adaptation is not able to make full use of the genetic repair effect to be discussed in Chapter 5. In addition, selection of mutation strengths is indirect and therefore highly disturbed, and Hansen [43] demonstrated that important invariance criteria do not hold if covariance matrices are adapted.

Cumulative mutation strength adaptation, introduced by Hansen and Ostermeier [42, 44], is an attempt to "derandomize" the process of strategy parameter adjustment. Unlike the previous two methods, this mutation strength adaptation scheme is deterministic rather than evolutive in that it explicitly analyzes statistical features of the selected offspring to drive the strategy parameter settings towards their optimal values. Instead of having differing strategy parameter settings compete with each other at a single time step, information from a number of time steps is accumulated and analyzed. Cumulative mutation strength adaptation has been demonstrated empirically to reliably adapt mutation covariance matrices on a variety of fitness landscapes. As we deal with isotropic mutations only, instead of outlining the mechanism in its full generality we restrict ourselves to the variant using a single mutation strength.

Cumulative mutation strength adaptation relies on the presumption that if the mutation strength is below its optimal value the selected consecutive steps tend to be parallel, and that if the mutation strength is too high consecutive steps tend to be antiparallel. So as to be able to reliably detect parallel or antiparallel correlations of progress vectors, information from a number of time steps needs to be accumulated. For the $(\mu/\mu, \lambda)$-ES, the *accumulated progress vector* $\mathbf{s}$ is defined by $\mathbf{s}^{(0)} = \mathbf{0}$ and the recursive relationship

$$\mathbf{s}^{(t+1)} = (1-c)\mathbf{s}^{(t)} + \sqrt{c(2-c)}\sqrt{\mu}\langle\mathbf{z}\rangle^{(t)}, \tag{2.3}$$

where c is a constant determining how far back the "memory" of the accumulation process reaches and where $\langle \mathbf{z} \rangle^{(t)} = \sum_{i=1}^{\mu} \mathbf{z}_{i;\lambda}^{(t)} / \mu$ is the progress vector defined in Equation (2.2). The mutation strength is updated according to

$$\sigma^{(t+1)} = \sigma^{(t)} \exp\left(\frac{\|\mathbf{s}^{(t+1)}\|^2 - N}{2DN} \right), \tag{2.4}$$

where D denotes a damping constant. Note that the term N in the numerator of the argument to the exponential function is the mean squared length of the accumulated progress vector if consecutive progress vectors are stochastically independent. If the squared length of the accumulated progress vector is less than N the mutation strength is decreased, if it is greater than N the mutation strength is increased. Also note that the prescription Equation (2.4) for adapting the mutation strength has been changed slightly from the prescription in the original algorithm given by Hansen [42] in that we perform adaptation on the basis of the squared length of the accumulated progress vector rather than on its length. Pending further investigation, the change has been approved of by Hansen and Ostermeier [44]. The constants c and D are set to $1/\sqrt{N}$ and $\sqrt{N}$, respectively, according to recommendations made by Hansen [42].

In Section 3 of Chapter 5, the performance of the $(\mu/\mu, \lambda)$-ES with cumulative mutation strength adaptation is analyzed. Given the inability of mutative self-adaptation to make full use of genetic repair, this seems like the more interesting mutation strength adaptation scheme to consider for the $(\mu/\mu, \lambda)$-ES. However, we also note that mutative self-adaptation may be appropriate for $(\mu/\rho \stackrel{+}{,} \lambda)$-ES with $\rho < \mu$, such as the $(\mu/2, \lambda)$-ES. Analyses of $(\mu/\rho \stackrel{+}{,} \lambda)$-ES with $1 < \rho < \mu$ are difficult and have not yet been attempted.

3. Fitness Environments

Throughout the present work, we assume that the determination of the fitness of a candidate solution is inexact due to noise in the evaluation or measurement procedure. Such *fitness noise* is commonly modeled by means of an additive, normally distributed term with mean zero. That is, when evaluating or measuring the fitness of a candidate solution $\mathbf{x} \in I\!R^N$, it is not the *ideal fitness* $f(\mathbf{x})$ that we obtain, but a *measured fitness* that is normally distributed with mean $f(\mathbf{x})$ and with standard deviation $\sigma_\epsilon(\mathbf{x})$. Quite naturally, $\sigma_\epsilon(\mathbf{x})$ is referred to as the *noise strength.* Fitness noise deceives the selection mechanism in that it can lead to inferior candidate solutions being selected based on their measured fitness while superior ones are discarded. For very high noise strength, selection is essentially random. Note that as an alternative or in addition to fitness noise, it may be of interest to consider the situation that it is the location of a candidate solution in search space that cannot be determined exactly. However, fitness noise is the predominant form of noise found in the literature and will be considered exclusively in what follows.

Fitness environments are characterized by a fitness function $f : I\!R^N \to I\!R$ and by a noise function $\sigma_\epsilon : I\!R^N \to I\!R$. So as to be able to analytically examine the performance of evolution strategies, only very simple fitness environments can be considered. Arguably the most simple nontrivial fitness environment possible is the one-dimensional, linear objective function

$$\begin{aligned} f_{line} &: I\!R \to I\!R \\ f_{line}(x) &= x \end{aligned} \tag{2.5}$$

with constant noise strength $\sigma_\epsilon(\mathbf{x}) = \sigma_\epsilon$. We will refer to this fitness environment as "the linear environment" and assume that the task at hand is maximization. Even in the linear environment, interesting insights can be gained. We will examine the performance of the (μ, λ)-ES in the linear environment in Chapter 4. Moreover, in a sense to be explained below, under certain simplifying assumptions the spherical environment to be introduced next can be reduced to the linear environment. Note that due to the scale invariance of the linear environment it is possible without loss of generality to assume that the strategy operates with unit mutation strength at *noise level* $\vartheta = \sigma_\epsilon/\sigma$. The linear environment is thus useless for the purpose of investigating optimal mutation strengths, where optimality of mutation strengths will be defined in Section 4.

The task that is studied the most commonly in connection with evolution strategies is minimization of the quadratic, spherically symmetric fitness function

$$\begin{aligned} f_{sphere}(\mathbf{x}) &: I\!R^N \to I\!R \\ f_{sphere}(\mathbf{x}) &= (\hat{\mathbf{x}} - \mathbf{x})^{\mathrm{T}}(\hat{\mathbf{x}} - \mathbf{x}) \end{aligned} \tag{2.6}$$

that maps candidate solution $\mathbf{x}$ to the square of its Euclidean distance $R = \|\hat{\mathbf{x}} - \mathbf{x}\|$ from the optimizer $\hat{\mathbf{x}}$. As the fitness function value of a candidate solution depends only on this distance, we sometimes write $f_{sphere}(R) = R^2$ for the fitness function. Following an argument by Rechenberg [66], the spherical environment can serve as a model for unconstrained optimization problems at a stage where the population of candidate solutions is already in relatively close vicinity to the optimizer. It derives part of its importance from that, ideally, mechanisms for the adaptation of mutation covariance matrices such as cumulative mutation strength adaptation rescale any convex-quadratic function into the sphere. Other fitness functions such as the ridge analyzed by Oyman et al. [61] attempt to model features of fitness landscapes farther away from the optimum.

Depending on the dependence of the noise strength on the location in search space, quite different behaviors of evolution strategies can be observed on the spherically symmetric fitness function. For example, if the noise strength is constant throughout the search space, it is impossible for an evolution strategy to accurately determine the location in search space of the optimizer. After much time has passed, the fitness values of the population of candidate solutions will fluctuate around a nonzero mean that increases with increasing noise strength and that can be decreased by increasing the population size. In what follows, however, we consider fitness-proportionate noise strength. More specifically, we assume that the noise strength for all offspring candidate solutions generated at time step t is proportional to the fitness $f(\langle \mathbf{x} \rangle^{(t)})$ of the centroid of the population at that time step. Such relative errors of measurement are of great practical importance as they arise for example in connection with physical measurement devices that are accurate up to a certain percentage of the quantity they measure. We will refer to the quadratic, spherically symmetric fitness function in connection with fitness-proportionate noise strength as "the spherical environment". Note that it would be somewhat more desirable to have the noise strength determined by the location in search space of the candidate solution being evaluated rather than by the centroid of the parental population. However, this latter noise model would considerably complicate the analysis. Moreover, we will see that in the limit of infinite search space dimensionality, the difference between the two noise models vanishes.

4. Measuring Performance

The local performance of an evolution strategy can be measured either in search space or in the space of fitness values. The corresponding performance measures are the *progress rate* and the *quality gain*, respectively. The quality gain is the expected difference in fitness of the population of candidate solutions at consecutive time steps. The progress rate is defined as the expected distance in search space traveled by the population in a particular direction within one time step. The direction is usually taken to be towards the optimizer if it exists, but has been defined alternatively by Rechenberg [67] as the direction of the local gradient of the fitness function. In the spherical environment both directions coincide. In the linear environment only the latter is well-defined. As we do not consider other fitness functions we feel justified in adopting Rechenberg's definition.

For the linear fitness environment the population consists of scalars x_i, $i = 1, \ldots, \mu$. We define the *one-generation gain* to be

$$\Delta_{line}^{(t)} = \frac{1}{\mu} \sum_{i=1}^{\mu} x_i^{(t+1)} - \frac{1}{\mu} \sum_{i=1}^{\mu} x_i^{(t)}. \tag{2.7}$$

Note that it does not make a difference whether we consider locations in search space of candidate solutions or their fitness values. Given the population of time step t, progress rate and quality gain are defined as the expected one-generation gain resulting from recombination, mutation, and selection. Note that even though strictly speaking we cannot talk of optimization in connection with f_{line} as an optimum does not exist, the local performance of evolution strategies in the linear environment is well defined in terms of the expected one-generation gain.

Clearly, the expected one-generation gain depends on the state of the population at time t. Depending on the type of strategy, this state is characterized by quantitics such as the distribution in search space of the candidate solutions, the degree of overvaluation of parental fitness to be introduced in Chapter 3, and the mutation strength. However, the dynamics of an evolution strategy in the linear environment are invariant to translations. Assuming constant mutation strength, after initialization effects have faded, quantities that characterize the population's state such as the central moments of the population as well as the expected one-generation gain of the strategy will have time-invariant statistical properties. It will be our goal to characterize this stationary state by means of making invariance assumptions.

In the spherical environment it does make a difference whether we measure the performance of an evolution strategy in search space or in the space of fitness values. Denoting the distance $\|\hat{\mathbf{x}} - \langle \mathbf{x} \rangle\|$ from the centroid of the population to the optimizer by R, the one-generation gain in search space is

$$\Delta_R^{(t)} = R^{(t)} - R^{(t+1)}$$

while the one-generation gain in the space of fitness values is

$$\begin{aligned} \Delta_{sphere}^{(t)} &= f\left(R^{(t)}\right) - f\left(R^{(t+1)}\right) \\ &= R^{(t)2} - R^{(t+1)2}. \end{aligned}$$

However, introducing normalized quantities

$$\Delta_R^* = \Delta_R \frac{N}{R} \qquad \text{and} \qquad \Delta_{sphere}^* = \Delta_{sphere} \frac{N}{2R^2}, \tag{2.8}$$

it follows by Taylor expansion that

$$\begin{aligned} \Delta_R^* &= N\left(1 - \sqrt{1 - \frac{2\Delta_{sphere}^*}{N}}\right) \\ &= \Delta_{sphere}^* + \frac{\Delta_{sphere}^{*\,2}}{2N} + \frac{\Delta_{sphere}^{*\,3}}{2N^2} + \frac{5\Delta_{sphere}^{*\,4}}{8N^3} + \ldots. \end{aligned}$$

It will become clear in the course of the following sections that the moments of Δ_{sphere}^* are of order unity in N. Taking expectations, on the right hand side all terms but the first vanish for $N \to \infty$. Thus, progress rate and quality gain as the expected values of the one-generation gains in search space and in the space of fitness values, respectively, are at least asymptotically equal. We will restrict ourselves to considering the quality gain in what follows and let "one-generation gain" refer to the one-generation gain in the space of fitness values.

As in the linear environment, progress rate and quality gain in the spherical environment depend on the state of the strategy at time t. Unlike the linear environment, the spherical environment does not exhibit invariance to translations but scale invariance. Introducing normalizations as in Equation (2.8), the normalized quantities describing the state of the strategy approach a time-invariant limit distribution provided that the mutation strength is adapted appropriately. As for the linear environment, our goal will be to characterize this limit distribution by making invariance assumptions, and to compute expected values of the quality gain. Moreover, so as to be able to talk of optimality of parameter settings, let us define the *efficiency* of an evolution strategy on the sphere as the normalized quality gain per evaluation of the fitness function

$$\eta = \frac{1}{\lambda}\mathrm{E}[\Delta^*_{sphere}]. \tag{2.9}$$

This definition clearly assumes that fitness function evaluations are the single dominant influence on the run time requirements of the evolution strategy and neglects for example the computational costs of recombination. We refer to those parameter settings of the $(\mu/\rho \overset{+}{,} \lambda)$-ES that maximize its efficiency as optimal.

5. Modeling the Sphere

In this section, we introduce a simple model for the behavior of evolution strategies in the spherical environment. We will make certain simplifications that are increasingly exact as the search space dimensionality increases. As the errors introduced by these simplifications vanish as N approaches infinity, we will refer to the model as "the sphere in the limit of infinite search space dimensionality". So as to verify that our analyses of the behavior of the model properly reflect features of the real system for finite N, it is necessary to check its accuracy by comparing the results of our calculations with measurements of the behavior of evolution strategies in the spherical environment. The simple model introduced in this section will be used throughout this work up to and including Section 1 of Chapter 5. In Section 2 of that chapter, a modification that greatly improves the accuracy of the results for strategies employing a nontrivial form of recombination will be introduced.

The analysis of the local performance of $(\mu/\rho \overset{+}{,} \lambda)$-ES on the sphere is greatly simplified by the symmetries inherent in both the strategies and the

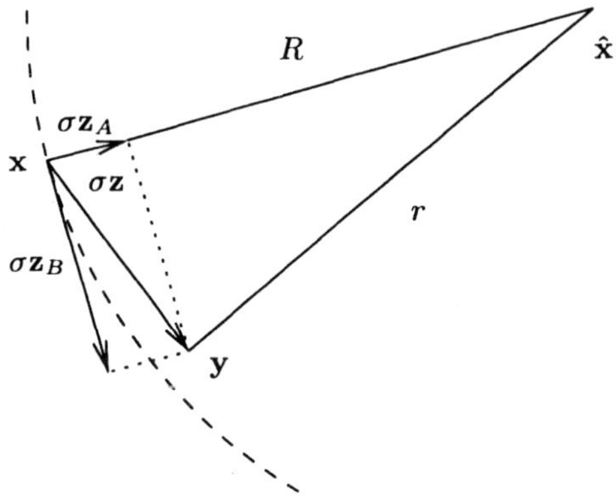

Figure 2.2. Decomposition of a vector $\mathbf{z}$ into central component $\mathbf{z}_A$ and lateral component $\mathbf{z}_B$ for the fitness function f_{sphere}. Vector $\mathbf{z}_A$ is parallel to $\hat{\mathbf{x}} - \mathbf{x}$, vector $\mathbf{z}_B$ is in the hyperplane perpendicular to that. The starting and end points, $\mathbf{x}$ and $\mathbf{y} = \mathbf{x} + \sigma\mathbf{z}$, of vector $\sigma\mathbf{z}$ are at distances R and r from the optimizer, respectively.

environment. A useful quantity to consider is the *fitness advantage* associated with a mutation vector which is defined as the difference

$$\delta = f(\mathbf{x}) - f(\mathbf{x} + \sigma\mathbf{z}) \tag{2.10}$$

between the ideal fitness values of starting point $\mathbf{x}$ and end point $\mathbf{x} + \sigma\mathbf{z}$, of the mutation. Following an idea put forward by both Beyer [18] and Rechenberg [67], a vector $\mathbf{z}$ originating at search space location $\mathbf{x}$ can be written as the sum of two vectors $\mathbf{z}_A$ and $\mathbf{z}_B$, where $\mathbf{z}_A$ is parallel to $\hat{\mathbf{x}} - \mathbf{x}$ and $\mathbf{z}_B$ is in the hyperplane normal to that. We will refer to $\mathbf{z}_A$ and $\mathbf{z}_B$ as the *central* and the *lateral components* of vector $\mathbf{z}$, respectively. We define the *signed length* z_A of the central component of vector $\mathbf{z}$ to equal $\|\mathbf{z}_A\|$ if $\mathbf{z}_A$ points towards the optimizer and to equal $-\|\mathbf{z}_A\|$ if $\mathbf{z}_A$ points away from it. Figure 2.2 illustrates the decomposition. Using elementary geometry and denoting the respective distances of $\mathbf{x}$ and $\mathbf{y} = \mathbf{x} + \sigma\mathbf{z}$ from the optimizer by R and r, it can be seen that

$$r^2 = (R - \sigma z_A)^2 + \sigma^2\|\mathbf{z}_B\|^2. \tag{2.11}$$

Introducing normalized quantities

$$\sigma^* = \sigma\frac{N}{R}, \qquad \delta^* = \delta\frac{N}{2R^2}, \qquad \text{and} \qquad \sigma_\epsilon^* = \sigma_\epsilon\frac{N}{2R^2}, \tag{2.12}$$

it follows from Equation (2.11) that the normalized fitness advantage associated with vector $\mathbf{z}$ is

$$\begin{aligned} \delta^* &= N\frac{R^2 - r^2}{2R^2} \\ &= \sigma^* z_A - \frac{\sigma^{*2}}{2N}\|\mathbf{z}\|^2. \end{aligned} \tag{2.13}$$

Note that the distance to the optimizer R does not appear explicitly on the right hand side of this equation. When observing the normalized quantity, the standard deviation of the noise term that stems from the error in the measurement of $f(\mathbf{x}+\sigma\mathbf{z})$ is σ_ϵ^*. On occasion, we will write $\delta_\epsilon^* = \delta^* + \sigma_\epsilon^* z_\epsilon$, where z_ϵ is drawn from a standardized normal distribution, for the normalized measured fitness advantage. Due to the assumption of fitness-proportionate noise strength, as the normalized fitness advantage, the normalized measured fitness advantage is independent of the location in search space. Thus, provided that the mutation strength σ is adapted such that the normalized mutation strength σ^* has a time-invariant distribution, we have scale invariance.

As mutations are isotropic we can without loss of generality assume that $\mathbf{z}_A = (z_1, 0, \ldots, 0)^{\mathrm{T}}$ and $\mathbf{z}_B = (0, z_2, \ldots, z_N)^{\mathrm{T}}$, where the z_i, $i = 1, \ldots, N$, are independently drawn from a standardized normal distribution. The squared length $\|\mathbf{z}\|^2$ of a mutation vector is the sum of the squared lengths z_A^2 and $\|\mathbf{z}_B\|^2$ of its central and lateral components. The squared length of the lateral component is the sum of squares of $N-1$ independent standardized normal variates and as such χ_{N-1}^2-distributed. According to Section 3.2 of Appendix A, as N tends to infinity, the distribution of the quotient $\|\mathbf{z}_B\|^2/N$ tends to normality with unity mean and its variance asymptotically scales as $2/N$. The squared length of the central component on the other hand is independent of N. Therefore, as N increases, the term $\|\mathbf{z}\|^2/N$ more and more closely approaches unity. In the limit of infinite search space dimensionality, the normalized fitness advantage associated with a mutation vector is

$$\delta^* = \sigma^* z_A - \frac{\sigma^{*2}}{2}. \tag{2.14}$$

The distribution of the normalized fitness advantage of mutation vectors thus approaches normality with mean $-\sigma^{*2}/2$ and with variance σ^{*2}. The asymptotic normality of the normalized fitness advantage has been shown formally employing an approach based on characteristic functions in [8]. Note that the contribution to the normalized fitness advantage of the central component of a mutation vector can be either positive or negative while the contribution of the lateral component is always negative. We will find the simple model given by Equation (2.14) sufficient to explain important properties of evolution strategies in search spaces of sufficiently high dimensionality.

We had stipulated in Section 3 that the noise strength at time step t be determined by the centroid of the parental population at that time, and we had noted that it would be more desirable to have the noise strength determined by the location in search space of the candidate solution being evaluated. Assuming the latter noise model, it follows from the assumption of fitness-proportionate noise strength that $\sigma_\epsilon(R)/R^2 = \sigma_\epsilon(r)/r^2$, and therefore that

$$\frac{\sigma_\epsilon(r)}{\sigma_\epsilon(R)} = 1 - \frac{2\delta^*}{N}.$$

As the normalized fitness advantage δ^* is of order unity in N, for $N \to \infty$ the difference between the noise strengths at the locations of parent and offspring tends to zero. Thus, in the limit of infinite search space dimensionality, the two noise models agree.

Chapter 3

THE $(1 + 1)$-ES: OVERVALUATION

It seems reasonable to start a discussion of the performance of evolution strategies in the presence of noise with the most simple strategy variants and then to proceed to increasingly more complex strategies. The $(1 + 1)$-ES is arguably the most simple evolution strategy. In its most basic form it is a simple stochastic hill climber. A parent $\mathbf{x}$ generates offspring candidate solution $\mathbf{x}+\sigma\mathbf{z}$ that replaces the parent in the next time step if and only if it appears superior in terms of its measured fitness. In this chapter, we analyze the local performance of the $(1 + 1)$-ES in the spherical environment in the limit of infinite search space dimensionality. We will see that the results provide a good understanding of the behavior of the $(1 + 1)$-ES on an N-dimensional sphere provided that N is not too small. The analysis of the local performance of the $(1 + 1)$-ES in the spherical environment is of particular interest as it is known from Beyer [27] that in the absence of noise its efficiency cannot be surpassed by any $(\mu/\mu, \lambda)$-ES. The local performance of the $(1 + 1)$-ES in the spherical environment has been analyzed by Beyer [18] under the assumption that the parental fitness is reevaluated along with that of the offspring in every time step. This assumption is likely not to hold for most implementations of the $(1 + 1)$-ES as it requires two rather than one fitness function evaluations per time step. In the following sections, the consequences of failure to reevaluate the parental fitness are explored. Rather than giving a full derivation of the results, we focus on the main steps and refer to Appendix C for details of the calculations.

1. Overvaluation

With selection based on measured fitness rather than on ideal fitness, a parental candidate solution with a measured fitness that is much higher than its ideal fitness is likely to persist for many time steps. Conversely, an offspring with a measured fitness that is lower than its ideal fitness is likely not to be accepted at all even though its ideal fitness may compare favorably with that of its parent. As a consequence, after several iterations of the mutation-selection mechanism the measured fitness of the parent is likely to be higher than its ideal fitness. We will see that this systematic overvaluation of the parental fitness has significant consequences for the local performance of the $(1+1)$-ES.

Throughout this chapter, let R and r denote the distances from the optimizer of parent and offspring at time t, respectively. Furthermore, let S^2 and s^2 denote the respective measured fitness of parent and offspring. The offspring is accepted if and only if $s^2 < S^2$. The difference $\Xi = R^2 - S^2$ between ideal fitness and measured fitness of the parent is in what follows referred to as the *degree of overvaluation* of the parental fitness. Likewise, the difference $\xi = r^2 - s^2$ between ideal fitness and measured fitness of the offspring is the degree of overvaluation of the offspring fitness. Due to the assumption of normally distributed, additive noise with mean zero and with variance $\sigma_\epsilon^2(R)$, the degree of overvaluation of the offspring fitness is normal with mean zero and with variance $\sigma_\epsilon^2(R)$. The degree of overvaluation of the parental fitness cannot as easily be given as it is a result of the repeated application of the mutation-selection mechanism.

Using the normalizations from Equation (2.12) and introducing degrees of normalized overvaluation

$$\Xi^* = \Xi\frac{N}{2R^2} \qquad \text{and} \qquad \xi^* = \xi\frac{N}{2R^2},$$

the normalized noise strength σ_ϵ^*, the degree of normalized overvaluation of the offspring ξ^*, and the normalized fitness advantage δ^* associated with mutation vectors are independent of the location in search space. In particular, the degree of normalized overvaluation of the offspring is normally distributed with mean zero and with variance σ_ϵ^{*2}. According to Section 5 of Chapter 2, the fitness advantage associated with a mutation vector is normally distributed with mean $-\sigma^{*2}/2$ and with variance σ^{*2}.

Let us for the moment assume that a mechanism that assures constancy of the normalized mutation strength σ^* is in place. We will discuss a mechanism for the adaptation of the mutation strength in Section 3.3. In that case, we have perfect scale invariance and the state of the $(1+1)$-ES is described by the degree of normalized overvaluation Ξ^* of the parental fitness. After initialization effects have faded, the distribution of Ξ^* converges towards a stable limit distribution. It is the goal of Section 1.1 to find an approximation to this distribution. In Section 1.2, several issues related to that distribution are discussed.

1.1 Obtaining the Limit Distribution

Let us attempt to derive an approximation to the stationary limit distribution of the normalized degree of overvaluation Ξ^*. For that purpose, we first determine the influence of mutation and selection on that distribution. We then determine characteristics of the distribution by means of requiring its stationarity. The approach is moment-based. More specifically, we proceed in three steps:

1 Expand the probability density function p_{Ξ^*} of the distribution of the degree of normalized overvaluation into a Gram-Charlier series. The expansion has the cumulants κ_i, $i \geq 1$, of the distribution as parameters that need to be determined. In practice, only the first k cumulants can be considered.

2 Compute the effects on the parental overvaluation of applying mutation and selection. In particular, the cumulants of the distribution of parental overvaluation after mutation and selection need to be determined.

3 Demand stationarity of the distribution. This amounts to requiring equality of the cumulants before and after mutation and selection and leads to a system of k equations in k unknowns, where k is the number of cumulants considered.

In practice, for the problem of determining the distribution of the degree of normalized parental overvaluation, neglecting all but the first two cumulants will turn out to yield very good results already. Effectively, this amounts to using a normal approximation

$$p_{\Xi^*}^{(t)}(x) = \frac{1}{\sqrt{\kappa_2}}\phi\left(\frac{x-\kappa_1}{\sqrt{\kappa_2}}\right), \tag{3.1}$$

where ϕ stands for the probability density function of the standardized normal distribution and has been defined in Section 3.1 of Appendix A. Note that Equation (3.1) is merely a model for the real distribution, and that we will run computer experiments to verify the accuracy of the model. We cannot give analytical bounds for the error introduced by omitting higher-order terms in the expansion as the shape of the distribution is entirely unknown.

Let $P_{\xi^*|\Xi^*}^{(acc)}(y|x)$ denote the chance of an offspring with degree of normalized overvaluation $\xi^* = y$ being accepted given a parent with degree of normalized overvaluation $\Xi^* = x$. Using the fact that the normalized fitness advantage δ^* associated with a mutation is normally distributed with mean $-\sigma^{*2}/2$ and with variance σ^{*2}, this probability can be computed as

$$\begin{aligned} P_{\xi^*|\Xi^*}^{(acc)}(y|x) &= \text{Prob}[s^2 < S^2] \\ &= \text{Prob}[r^2 - \xi < R^2 - \Xi] \\ &= \text{Prob}[x - y < \delta^*] \\ &= \Phi\left(\frac{y - x - \sigma^{*2}/2}{\sigma^*}\right), \end{aligned} \tag{3.2}$$

where Φ stands for the distribution function of the standardized normal distribution.

At time $t+1$, the degree of normalized overvaluation of the parental fitness equals $\Xi^{*(t)}$ if the offspring has been rejected, and it equals ξ^* if the offspring has been accepted. Thus, the degree of normalized overvaluation of the parental fitness at time $t+1$ is in an interval of width $\mathrm{d}x$ centered at x if either the degree of normalized overvaluation of the offspring is in that interval and the offspring is accepted, or if the degree of normalized overvaluation of the parent is in that interval and the offspring is rejected. Therefore, the probability density function of the degree of normalized overvaluation of the parental fitness at time $t+1$ is

$$p_{\Xi^*}^{(t+1)}(x) = \int_{-\infty}^{\infty} p_{\Xi^*}^{(t)}(y) p_{\xi^*}(x) P_{\xi^*|\Xi^*}^{(acc)}(x|y)\mathrm{d}y + \int_{-\infty}^{\infty} p_{\Xi^*}^{(t)}(x) p_{\xi^*}(y) \left[1 - P_{\xi^*|\Xi^*}^{(acc)}(y|x)\right] \mathrm{d}y, \quad (3.3)$$

where the first summand results from offspring candidate solutions that are accepted while the second term reflects parental candidate solutions that are retained. Probability density $p_{\Xi^*}^{(t)}$ is given by Equation (3.1), probability density p_{ξ^*} is that of a normal distribution with mean zero and with variance σ_ϵ^{*2}. Stationarity requires that

$$\mathrm{E}\left[\Xi^{*(t+1)}\right] = \kappa_1 \quad (3.4)$$

and that

$$\mathrm{Var}\left[\Xi^{*(t+1)}\right] = \kappa_2. \quad (3.5)$$

Equations (3.4) and (3.5) form a system of two nonlinear equations that need to be solved for the two unknowns κ_1 and κ_2.

Determining the probability density $p_{\Xi^*}^{(t+1)}$ and the mean and the variance thereof is cumbersome but straightforward. Detailed calculations can be found in Section 2 of Appendix C. Letting $\chi^{(k)}$ denote the kth derivative with respect to κ_1 of

$$\chi(\kappa_1, \kappa_2) = \Phi\left(\frac{\sigma^{*2}/2 + \kappa_1}{\sqrt{\sigma^{*2} + \sigma_\epsilon^{*2} + \kappa_2}}\right), \quad (3.6)$$

then, according to Equations (C.5) and (C.6) and introducing symbols g_1 and g_2 for future reference, mean and variance of $\Xi^{*(t+1)}$ are

$$\begin{aligned} \mathrm{E}\left[\Xi^{*(t+1)}\right] &= g_1(\kappa_1, \kappa_2) \\ &= \kappa_1 \chi^{(0)}(\kappa_1, \kappa_2) + \left(\sigma_\epsilon^{*2} + \kappa_2\right) \chi^{(1)}(\kappa_1, \kappa_2) \end{aligned} \quad (3.7)$$

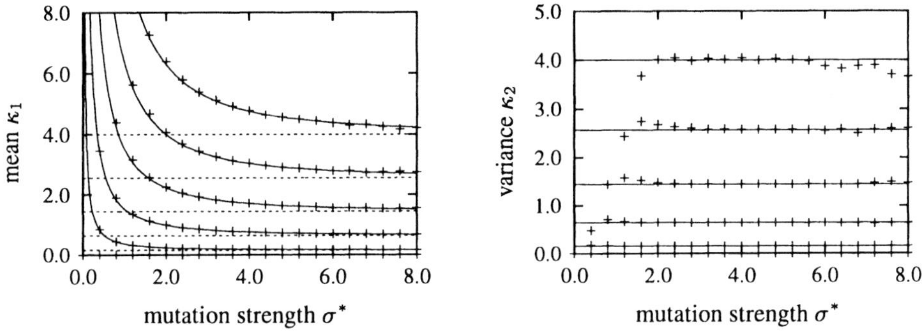

Figure 3.1. Mean κ_1 and variance κ_2 of the degree of normalized overvaluation Ξ^* of the parental fitness as functions of normalized mutation strength σ^*. The solid curves display the results of numerical root finding of Equation (3.11) and of Equation (3.10), respectively, for, from top to bottom, normalized noise strengths $\sigma_\epsilon^* = 2.0, 1.6, 1.2, 0.8, 0.4$, and 0.0. The crosses represent data measured in experiments described in the text. The dashed lines in the left hand graph indicate the limit values for high normalized mutation strength derived in Section 1.2.

and

$$\begin{aligned}\text{Var}\left[\Xi^{*(t+1)}\right] &= g_2(\kappa_1, \kappa_2) \\ &= \left(\sigma_\epsilon^{*2} - \kappa_2\right)\left(1 - \frac{\partial g_1}{\partial \kappa_1}\right) \\ &\quad + g_1(\kappa_1, \kappa_2)\left(\kappa_1 - g_1(\kappa_1, \kappa_2)\right) + \kappa_2, \qquad (3.8)\end{aligned}$$

respectively. Partial derivatives of χ are given in Section 1 of Appendix C.

Equations (3.7) and (3.8) form a two-dimensional iterated map describing the change from one time step to the next of the parameters κ_1 and κ_2 that determine the distribution of the degree of normalized overvaluation. After sufficiently many time steps, convergence to a fixed point (κ_1, κ_2) can be observed. To find the fixed point, inserting Equation (3.8) into Equation (3.5) while making use of Equation (3.4) yields condition

$$\left(\sigma_\epsilon^{*2} - \kappa_2\right)\left(1 - \left.\frac{\partial g_1}{\partial \kappa_1}\right|_{(\kappa_1, \kappa_2)}\right) = 0 \qquad (3.9)$$

Clearly, Equation (3.9) is satisfied for

$$\kappa_2 = \sigma_\epsilon^{*2}. \qquad (3.10)$$

Inserting Equation (3.7) into Equation (3.4) and using Equation (3.10), the other condition reads

$$\chi^{(1)}\left(\kappa_1, \sigma_\epsilon^{*2}\right) = \frac{\kappa_1}{2\sigma_\epsilon^{*2}}\left[1 - \chi^{(0)}\left(\kappa_1, \sigma_\epsilon^{*2}\right)\right]. \qquad (3.11)$$

The result of numerical root finding for κ_1 as well as the solution for κ_2 from Equation (3.10) are displayed in Figure 3.1. Also displayed are the results of computer experiments in which the degree of normalized overvaluation of the parental fitness Ξ^* is initialized by generating a random sample from a normal distribution with mean κ_1 and with variance κ_2, where κ_1 and κ_2 are determined from Equations (3.10) and (3.11). Then, for 10^9 time steps, the normalized fitness advantage δ^* and the degree of normalized overvaluation of an offspring ξ^* are generated by randomly sampling from normal distributions with mean $-\sigma^{*2}/2$ and variance σ^{*2} and with mean zero and variance σ_ϵ^{*2}, respectively. If $\Xi^* < \delta^* + \xi^*$, then the offspring is accepted and Ξ^* is replaced by ξ^*. Otherwise, Ξ^* remains unchanged. Mean and variance of Ξ^*, the success probability to be introduced in Section 2.1, and the quality gain are obtained by averaging over all time steps. As we will see in Section 2.1, for nonzero noise, success probabilities can be very small for large and in particular for small mutation strengths. As a consequence, for such parameter settings, stagnation times can be very long compared to the number of time steps the experiments are run for, and averaging will not yield good estimates for the mean and the variance of the degree of normalized overvaluation. Such unsatisfactory results have been omitted from Figure 3.1. We consider the good agreement between the experiments and the results obtained on the basis of the normal approximation as evidence for the accuracy of the approach.

1.2 Limit Behavior, Stability, and Convergence Rates

This section discusses several issues related to the limit distribution of the degree of normalized overvaluation found in the previous section. In particular, the behavior of the mean degree of normalized overvaluation in the limit of very large normalized mutation strengths is analyzed, the stability of the limit distribution is investigated, and the rate of convergence with which the limit distribution is approached is studied.

Limit behavior

For finite normalized noise strength σ_ϵ^*, the expected degree of normalized overvaluation κ_1 approaches a finite constant value as the normalized mutation strength σ^* is increased. Using de l'Hôspital's rule, Equations (C.1) and (C.3), and $\lim_{\sigma^*\to\infty} \partial\kappa_1/\partial\sigma^* = \partial\kappa_2/\partial\sigma^* = 0$, it follows that

$$\begin{aligned}
\lim_{\sigma^*\to\infty} \frac{\chi^{(1)}(\kappa_1,\kappa_2)}{1-\chi^{(0)}(\kappa_1,\kappa_2)} &= \lim_{\sigma^*\to\infty} -\frac{\chi^{(2)}(\kappa_1,\kappa_2)+\chi^{(3)}(\kappa_1,\kappa_2)}{\chi^{(1)}(\kappa_1,\kappa_2)+\chi^{(2)}(\kappa_1,\kappa_2)} \\
&= \lim_{\sigma^*\to\infty} \frac{\sigma^{*2}/2-\sigma^{*2}/4}{\sigma^{*2}-\sigma^{*2}/2} \\
&= 1/2.
\end{aligned}$$

Thus, Equation (3.11) shows that

$$\lim_{\sigma^* \to \infty} \kappa_1 = \sigma_\epsilon^{*2}. \tag{3.12}$$

These limit values are shown as dashed lines in the left hand graph of Figure 3.1.

Stability

Section 1.1 has shown the existence of a fixed point of the iterated map (g_1, g_2) given by Equations (3.7) and (3.8). The stability of this fixed point can be shown by linearizing the system at the location of the fixed point and showing that the resulting map is volume-contracting. For that purpose, the eigenvalues η_1 and η_2 of the Jacobian matrix

$$M = \begin{pmatrix} \dfrac{\partial g_1}{\partial \kappa_1} & \dfrac{\partial g_1}{\partial \kappa_2} \\ \dfrac{\partial g_2}{\partial \kappa_1} & \dfrac{\partial g_2}{\partial \kappa_2} \end{pmatrix} \tag{3.13}$$

at the fixed point determined by Equations (3.10) and (3.11) have to be shown to be less than 1.0 in absolute value. For an introduction to stability theory of iterated maps we refer to Luenberger [54]. Computation of the eigenvalues is carried out in Section 3 of Appendix C. The results are displayed in Figure 3.2. It can be observed that the eigenvalues of the system are indeed less than 1.0 in absolute value for any normalized mutation strength σ^*, $0 < \sigma^* < \infty$. Furthermore, they are always real, showing that the approach of the fixed point does not involve oscillatory behavior. For nonzero noise, both eigenvalues tend to 1.0 for both $\sigma^* \to 0$ and $\sigma^* \to \infty$. Thus, for very small and for very large normalized mutation strengths the fixed point is almost neutral. A reason for that behavior will become obvious in the discussion of success probabilities in Section 2.1.

Convergence rates

While the behavior of the degree of normalized overvaluation after long time spans has been investigated in the previous sections, nothing has been said regarding the amount of time required to arrive at behavior which is statistically indistinguishable from that limit behavior. Numerical experiments indicate that this time can be very long especially for small normalized mutation strengths. For $\sigma^* = 0$ and nonzero noise strength, at time step t, the degree of normalized overvaluation of the parental fitness is the maximum of t independent, normally distributed random variables with mean zero and with variance σ_ϵ^{*2}. Thus, the behavior in time of the degree of normalized overvaluation is governed by order statistics. Using a bound derived by Beyer [27], p.77, the expected value of the first order statistic of t independent samples of a random variable with

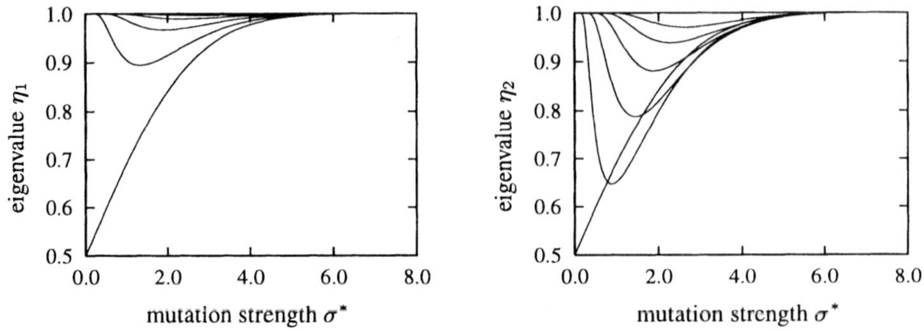

Figure 3.2. Eigenvalues η_1 and η_2 of the Jacobian given in Equation (3.13) as functions of the normalized mutation strength σ^*. The curves display, from top to bottom, the results for noise strengths $\sigma_\epsilon^* = 2.0, 1.6, 1.2, 0.8, 0.4$, and 0.0.

standardized normal distribution is of order $\sqrt{\log t}$. Thus, for $\sigma^* = 0$, the growth of the mean degree of normalized overvaluation is sublogarithmic. A very slow approach of the limit distribution can also be observed for small, nonzero normalized mutation strength.

However, it is also true that the initial approach of the fixed point is rather fast in the range of mutation strengths which will in Section 2 be seen to be the range of interest. Unless the mutation strength is chosen very high, a considerable portion of the final overvaluation builds up within the first few time steps already. Thus, it can be expected that the effects of overvaluation become observable rather early during the evolutionary search.

2. Performance

This section discusses the performance of the $(1+1)$-ES on the noisy sphere. In Section 2.1, the success probability is introduced, offering an explanation for the observations regarding the stability of the fixed point made in the previous section. In Section 2.2, the normalized quality gain is computed, and in Section 2.3 optimal parameter settings are obtained.

2.1 Success Probabilities

The *success probability* P_{succ} is the probability with which a parent is replaced by an offspring. It can be computed from the conditional probability $P^{(acc)}_{\xi^*|\Xi^*}$ defined in Section 1.1 as

$$P_{succ} = \int_{-\infty}^{\infty} p_{\Xi^*}(x) \int_{-\infty}^{\infty} p_{\xi^*}(y) P^{(acc)}_{\xi^*|\Xi^*}(y|x)\mathrm{d}y\mathrm{d}x. \tag{3.14}$$

Evaluation of the integrals is once again straightforward. Detailed calculations can be found in Section 4.1 of Appendix C. According to Equation (C.8), the

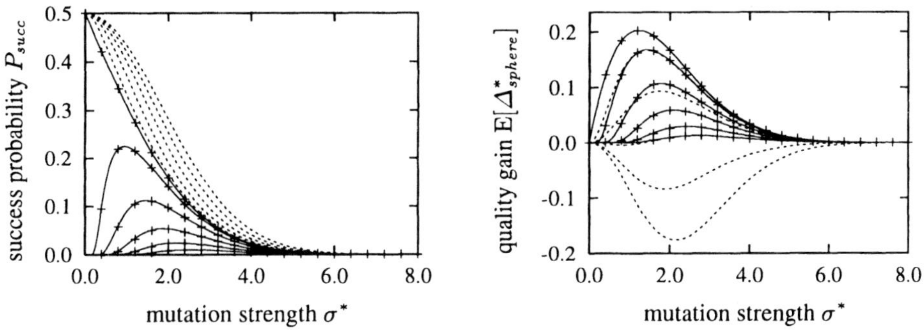

Figure 3.3. Success probability P_{succ} and normalized quality gain $E[\Delta^*_{sphere}]$ as functions of normalized mutation strength σ^*. The solid curves display, from top to bottom, the results from Equations (3.15) and (3.18) for normalized noise strengths $\sigma_\epsilon^* = 0.0, 0.4, 0.8, 1.2, 1.6$, and 2.0. The crosses represent data measured in experiments described in Section 1.1. The dashed curves display the corresponding results from Beyer [18] for a strategy that reevaluates the fitness of the parent at every time step. Note that in the left hand graph, for the latter strategy the noise strength increases from bottom to top.

success probability reads

$$P_{succ} = 1 - \chi^{(0)}(\kappa_1, \kappa_2) \,. \tag{3.15}$$

With the values for κ_1 and κ_2 obtained from Equations (3.10) and (3.11), in Figure 3.3 it is compared with measurements from the computer experiments described in Section 1.1. The agreement is excellent, demonstrating the quality of the normal assumption made in Section 1.1. Moreover, in Figure 3.4, the results are compared with measurements of a $(1+1)$-ES on a noisy sphere with search space dimensionality $N = 40$. It can be seen that the predictions afforded by Equation (3.15) are quite accurate for that moderate finite value of N even though they have been obtained in the limit of infinite search space dimensionality.

The reciprocal quantity $1/P_{succ}$ of the success probability is the average number of time steps that a candidate solution survives. As for nonzero noise strength the success probability approaches zero both for small and for large normalized mutation strengths, long periods of stagnation can be observed for such parameter settings. This explains the approximate neutrality of the fixed point observed in Section 1.2.

Comparing the dependence of the success probability on the normalized mutation strength, qualitative differences between the $(1+1)$-ES with and without reevaluation of the parental fitness can be observed. Results arrived at by Beyer [18] for a $(1+1)$-ES with reevaluation are included in Figure 3.3 for reference. Alternatively, the same results can be arrived at from Equation (3.15) with $\kappa_1 = 0$ as reevaluating the parental fitness yields a degree of normalized overvaluation of the parental fitness which is normal with mean zero and with

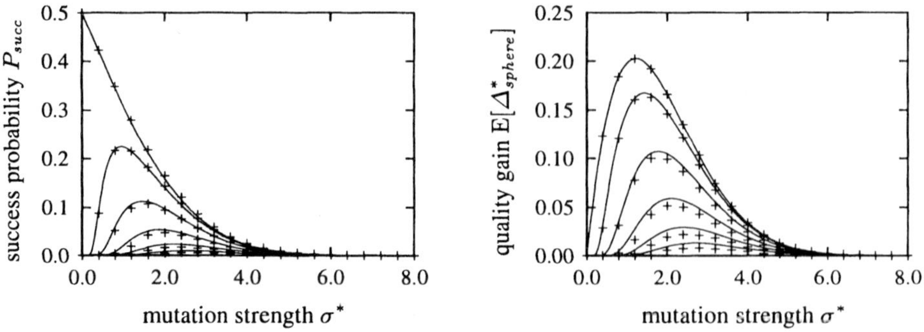

Figure 3.4. Success probability P_{succ} and normalized quality gain $E[\Delta^*_{sphere}]$ of the $(1+1)$-ES on the noisy sphere with search space dimensionality $N = 40$ as functions of normalized mutation strength σ^*. The crosses mark data obtained in computer experiments with, from top to bottom, normalized noise strengths $\sigma^*_\epsilon = 0.0, 0.4, 0.8, 1.2, 1.6$, and 2.0. The solid lines represent the corresponding predictions from Equations (3.15) and (3.18) obtained in the limit of infinite search space dimensionality.

variance σ^{*2}_ϵ. In general, with reevaluation of the parental fitness, increasing the noise strength increases the success probability; without reevaluation, it decreases it. The difference between the two strategies is particularly pronounced for small normalized mutation strength. While with reevaluation the chance of accepting an offspring is close to one half, that chance is close to zero — except in the absence of noise — if the parental fitness is not reevaluated. With reevaluation of the parental fitness, for high noise strength and small mutation strength, the strategy performs a random walk in search space. Without reevaluation, the search simply stagnates.

2.2 Quality Gain

The normalized one-generation gain Δ^*_{sphere} equals zero if the offspring is rejected, and it equals the normalized fitness advantage δ^* of the offspring if it is accepted. The normalized quality gain is thus

$$E[\Delta^*_{sphere}] = \int_{-\infty}^{\infty} p_{\Xi^*}(x) \int_{-\infty}^{\infty} y p_{\delta^*}(y) P^{(acc)}_{\delta^*|\Xi^*}(y|x)\mathrm{d}y\mathrm{d}x, \tag{3.16}$$

where, using the fact that ξ^* is normally distributed with mean zero and with variance σ^{*2}_ϵ, the probability of accepting an offspring with a normalized fitness advantage $\delta^* = y$ given a parent with degree of normalized overvaluation $\Xi^* = x$ is

$$\begin{aligned} P^{(acc)}_{\delta^*|\Xi^*}(y|x) &= \mathrm{Prob}(s^2 < S^2) \\ &= \mathrm{Prob}(r^2 - \xi < R^2 - \Xi) \end{aligned}$$

$$= \text{Prob}(x - y < \xi^*)$$
$$= \Phi\left(\frac{y-x}{\sigma_\epsilon^*}\right). \tag{3.17}$$

Solving the integrals in Equation (3.16) is straightforward. Detailed calculations can be found in Section 4.2 of Appendix C. According to Equation (C.9), the result reads

$$\text{E}[\Delta^*_{sphere}] = -\frac{\sigma^{*2}}{2}\left[1 - \chi^{(0)}(\kappa_1, \kappa_2)\right] + \sigma^{*2}\chi^{(1)}(\kappa_1, \kappa_2). \tag{3.18}$$

In Figure 3.3, it is compared with measurements from the computer experiments described in Section 1.1. The values for κ_1 and κ_2 have been obtained from Equations (3.10) and (3.11). The agreement is excellent, demonstrating the quality of the normal approximation made in Section 1.1. Moreover, in Figure 3.4, the results are compared with measurements of a $(1+1)$-ES on a noisy sphere with search space dimensionality $N = 40$. It can be seen that the predictions afforded by Equation (3.18) that has been derived in the limit of infinite search space dimensionality are quite accurate even for moderate finite value of N.

When comparing the dependence of the normalized quality gain on the normalized mutation strength, qualitative differences between the $(1+1)$-ES with and without reevaluation of the parental fitness can be observed. Results arrived at by Beyer [18] for a $(1+1)$-ES with reevaluation are included in Figure 3.3 for reference. The same results can be obtained from Equation (3.18) with $\kappa_1 = 0$ as explained in Section 2.1. With reevaluation of the parental fitness, the normalized quality gain is negative for a wide range of mutation strengths if the normalized noise strength exceeds a value of about 1.0. For σ_ϵ^* greater than about 1.4, the normalized quality gain is negative except for very high normalized mutation strengths. Without reevaluation on the other hand, the normalized quality gain is non-negative for any normalized mutation strength.

2.3 Optimal Parameter Settings

The performance of the $(1+1)$-ES for optimally chosen normalized mutation strength, i.e. for the normalized mutation strength that maximizes the normalized quality gain, is of particular interest. It can be obtained from Equation (3.18) by computing the derivative of the normalized quality gain with respect to σ^* and finding a root thereof. Figure 3.5 displays the success probability and the normalized quality gain for optimally chosen normalized mutation strength for both the strategies with and without reevaluation of the parental fitness. The right hand graph shows that the strategy without reevaluation is never inferior, but is clearly superior to that with reevaluation for high noise strengths. In that case, the strategy with reevaluation requires very high mutation strengths

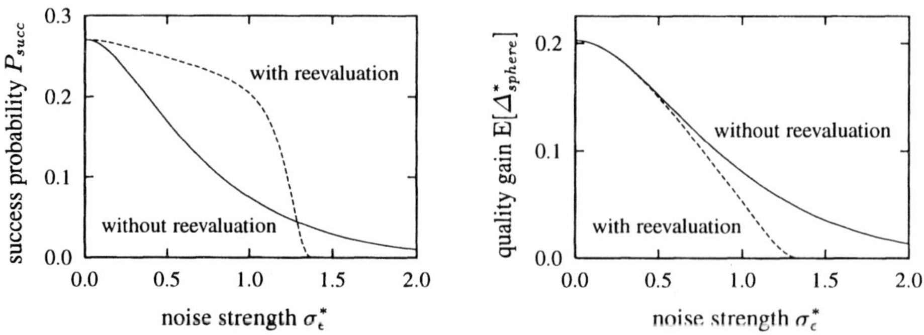

Figure 3.5. Success probability P_{succ} and normalized quality gain $E[\Delta^*_{sphere}]$ in the case of optimally chosen normalized mutation strength as functions of normalized noise strength σ_ϵ^*. The solid and the dashed curves display the results for the strategies without and with reevaluation of the parental fitness, respectively.

to effectively reduce the success probability to zero so as to avoid negative normalized quality gain. The strategy without reevaluation on the other hand is still capable of producing measurable positive normalized quality gain. Considering efficiency, i.e. quality gain per evaluation of the fitness function, the strategy without reevaluation fares even better as it requires only one fitness function evaluation per time step, as compared to the two required for the strategy with reevaluation.

3. Discussion

In this section we discuss the benefits of a positive degree of normalized overvaluation of the parental fitness, the use of resampling to improve the efficiency of the $(1+1)$-ES in the presence of noise, and the problem of adapting the mutation strength based on observed success probabilities.

3.1 Overvaluation Revisited

Section 2.3 has shown that a $(1+1)$-ES which does not reevaluate the parental fitness locally outperforms one which does on the noisy quadratic sphere in the limit of infinite search space dimensionality. Failure to reevaluate the parental fitness has been seen to lead to systematic overvaluation of the parental fitness. For zero mutation strength, it has been shown that the degree of normalized overvaluation of the parental fitness grows sublogarithmically with time. For nonzero mutation strength, from time to time a gain in ideal fitness acts to reduce the degree of normalized overvaluation, which in turn approaches a stable limit distribution.

The reason for the improved performance of the strategy without reevaluation are the reduced success probabilities that are a consequence of systematic overvaluation of the parental fitness. A small gain in measured fitness is often due to noise rather than to a gain in ideal fitness. Positive overvaluation raises the bar for an offspring candidate solution to be accepted, preventing the strategy from making steps purely due to noise that are likely to lead it away from the optimizer.

However, positive overvaluation also leads to the rejection of offspring candidate solutions which are inferior in terms of measured fitness but superior in terms of ideal fitness and which would be accepted if no overvaluation were present. Clearly, both zero overvaluation and infinite overvaluation are less than optimal. It is interesting to ask what degree of normalized overvaluation would be optimal if it could be selected deliberately instead of being a result of the interplay between mutation and selection. As, using Equation (3.18),

$$\frac{\partial}{\partial \kappa_1} \mathrm{E}[\Delta^*_{sphere}] = \sigma^{*2} \left(\frac{1}{2} \chi^{(1)} + \chi^{(2)} \right),$$

optimal κ_1 necessarily requires $\chi^{(1)}/2 + \chi^{(2)} = 0$ and therefore, using Equation (C.1), $\kappa_1 = \sigma_\epsilon^{*2}$. This value agrees exactly with the degree of normalized overvaluation that is obtained for $\sigma^* \to \infty$. For finite normalized mutation strength however, as can be seen from Figure 3.1, the expected degree of normalized overvaluation that results from the interplay between mutation and selection is higher than optimal. Therefore, occasional reevaluation of the parental fitness may be preferable to both no reevaluation at all and to reevaluation in every time step.

3.2 The Use of Resampling

In some situations, the accuracy of a fitness function evaluation depends on the time devoted to it. For example, fitness evaluation may involve the solution of a partial differential equation, where the accuracy depends on the grid size used. Another example is the evaluation of the performance of a neural network, where the accuracy can be improved by increasing the number of input patterns the network is run on. A third example occurs when fitness evaluation involves running a simulation, and if additional accuracy is gained when the simulation time span is increased.

Of interest here is the case that fitness evaluation involves some physical measurement that is performed with limited accuracy. In that case, the accuracy of the fitness estimate can be increased by averaging over a number of independent measurements. Averaging over k independent evaluations of the fitness function at any one search space location effectively reduces the noise strength to $\sigma_\epsilon/\sqrt{k}$, albeit at the cost of a k-fold increase in computational costs even if the overhead resulting from averaging is ignored. For given normalized noise strength σ_ϵ^*, the efficiency of the $(1+1)$-ES with k-times sampling is

$$\eta_k(\sigma_\epsilon^*) = \frac{1}{k} \max_{\sigma^* \in I\!R_+} \left(\mathrm{E}[\Delta^*_{sphere}] \big|_{\sigma_\epsilon^*/\sqrt{k}} \right).$$

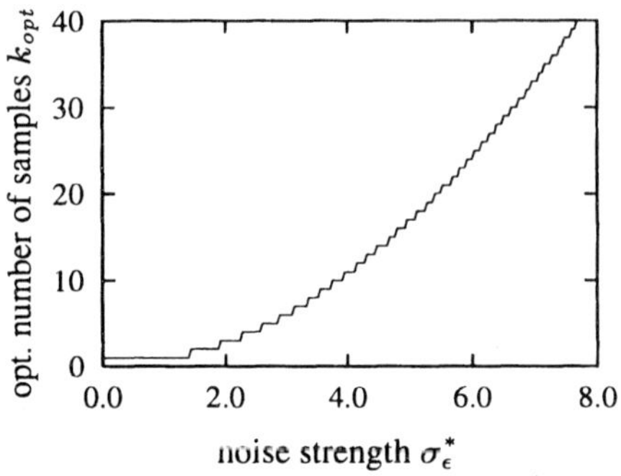

Figure 3.6. Optimal number k_{opt} of fitness function evaluations to be averaged over per time step as a function of normalized noise strength σ_ϵ^*.

For any given normalized noise strength σ_ϵ^* there is a $k = k_{opt}$ which maximizes $\eta_k(\sigma_\epsilon^*)$. Figure 3.6 shows values of k_{opt} that have been obtained numerically for a range of normalized noise strengths for a strategy which does not reevaluate the parental fitness. It can be seen that the $(1+1)$-ES is capable of operating efficiently only up to a normalized noise strength of about 1.4. For higher normalized noise strengths, explicit averaging of multiple fitness values is recommended. However, we will see in Chapters 4 and 5 that in the presence of noise, other types of evolution strategy can far exceed the efficiency of the $(1+1)$-ES.

3.3 Mutation Strength Adaptation

The analysis of the preceding sections relies on the scale invariance of the sphere and therefore on the constancy of the normalized mutation strength σ^*. Due to the fact that the distance R to the optimizer changes, a mechanism for the adaptation of the mutation strength σ is required. In Section 2 of Chapter 2, we outlined the one-fifth success rule for that purpose. It has to be noted that originally, this rule was suggested only for the noise-free case. However, as shown by Beyer [18] and also as plausible from the left hand graph of Figure 3.5 in combination with Figure 3.3, the rule is useful for the noisy sphere as well if the normalized noise strength does not exceed a value of about 1.2 and if the parental fitness is reevaluated in every time step. However, the figures also suggest that it is not suitable for strategies that do not reevaluate parental fitness except for relatively moderate normalized noise strengths. For normalized noise strengths exceeding a value of about 0.47, a success probability of 0.2 is not achieved for any mutation strength. Even worse, it is not true that decreasing the mutation strength always increases the success probability and vice versa. As a result, after a number of mutation strength adaptations, the mutation strength will be so small that the normalized mutation strength is in the regime to the left of the maximum in Figure 3.3. Further reductions of the mutation strength act to further reduce the success probability and therefore lead in turn to a further reduction of the mutation strength. After a while, the decay of the normalized mutation strength becomes essentially exponential.

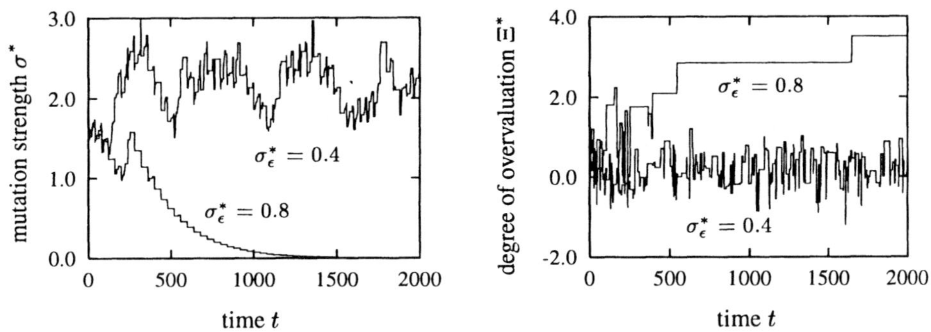

Figure 3.7. Results of two typical runs of a $(1+1)$-ES with mutation strength adaptation based on observed success probabilities. Parental fitness values are not reevaluated. Normalized noise strengths are $\sigma_\epsilon^* = 0.4$ and $\sigma_\epsilon^* = 0.8$. The left hand graph shows the development of the normalized mutation strength σ^* over time t, the right hand graph that of the degree of normalized overvaluation Ξ^*.

Figure 3.7 illustrates that this effect is not merely academic but that it is observable also in finite-dimensional search spaces and on relatively short time scales. It shows results from two typical runs of a $(1+1)$-ES using a mutation strength adaptation mechanism that differs from the one suggested by Schwefel that was quoted in Section 3 of Chapter 2 only in that an estimate of the success probability is obtained by averaging over N time steps rather than over $10N$. The change has been made so as to be able to adapt the mutation strength earlier during the evolution and is rather insignificant for the performance of the algorithm on the sphere. The objective function is a quadratic sphere with search space dimensionality $N = 40$. The strategy does not reevaluate the parental fitness. Experiments have been conducted with normalized noise strengths $\sigma_\epsilon^* = 0.4$ and $\sigma_\epsilon^* = 0.8$. For $\sigma_\epsilon^* = 0.4$, the mutation strength adaptation rule works reasonably well and achieves an average normalized mutation strength which ensures continued progress. For $\sigma_\epsilon^* = 0.8$ however, the normalized mutation strength tends to zero after a number of mutation strength adaptations. After time step 500, in only two time steps a positive gain of ideal fitness is achieved. The right hand graph of Figure 3.7 shows that overvaluation of the parental fitness which has been identified as the underlying reason for the stagnation is indeed present as expected. While the limit state is not reached within $2,000$ time steps, overvaluation already has a significant impact on the performance of the strategy.

Note that the mutation strength adaptation rule based on success probabilities cannot easily be fixed to work in the presence of noise merely by reducing the target success probability from 0.2 to lower values as has been suggested by Beyer [18] for the strategy that reevaluates parental fitness. First of all, the optimal success probability is not easily obtained. Moreover, in the presence of noise it is not unique. As has been noted above, the adaptation scheme relies on the monotonicity of the dependence of the success probability on the mutation strength. Second, in the presence of noise, the success probability is not easily measured. In runs of evolution strategies, long periods of stagnation (if the parent is highly overvalued) change with bursts of activity (if the overvaluation of the parent is rather low). Compiling the number of successes over the most recent time steps does not usually yield a good estimate of the success probability. Finally, even if a success probability-based mutation strength adaptation scheme that works on the sphere could be found, it is unclear whether it would work in other fitness environments as well. The one-fifth success rule derives its attractiveness from the fact that it has been found to work on a variety of fitness functions in the absence of noise.

To summarize, great care has to be taken if a success probability-based mutation strength adaptation rule is used in a noisy environment if measured fitness values can survive for more than a single time step. The presence of systematic overvaluation can lead to the violation of the assumptions on which rules such as the one-fifth success rule rest. Of course, averaging over a number of samples can always be used so as to reduce the noise strength enough to make success probability-based rules work, but neither is it easy to determine whether a particular resampling rate is sufficient, nor do the computational costs involved make this choice very appealing. While overvaluation has been shown to have the potential to beneficially affect the local performance of the $(1+1)$-ES, reevaluation of the parental fitness may be necessary to make success probability-based mutation strength adaptation viable.

Chapter 4

THE (μ, λ)-ES: DISTRIBUTED POPULATIONS

Analyzing the local performance of the (μ, λ)-ES is much more difficult than analyzing that of the $(1+1)$-ES. While for the $(1+1)$-ES the population consists of a single candidate solution, for the (μ, λ)-ES with $\mu > 1$ the population of candidate solutions that emerges in the course of evolution is distributed in search space and needs to be modeled. In the absence of noise, Beyer [20] presented a moment-based analysis of the performance of the (μ, λ)-ES for spherically symmetric fitness function. The approach was based on expanding the distribution of offspring candidate solutions in terms of derivatives of the normal distribution. Approximations to the lower-order central moments of the distribution and subsequently to the progress rate were obtained by imposing "self-consistency conditions" and solving a resulting system of equations. The results that were obtained are quite accurate even if only variance and skewness of the distribution are considered. A main result of Beyer's analysis, which was also stated by Rechenberg [67], is the observation that on the noise-free sphere the performance of the (μ, λ)-ES with $\mu > 1$ is never superior to that of the $(1, \lambda)$-ES, and that thus no benefits can be gained from retaining any but the best candidate solution generated. However, Rechenberg also provided empirical evidence that this is not true in the presence of noise. Simple computer experiments can be used to demonstrate that for the very same fitness function, significant speed-up factors over the $(1, \lambda)$-ES can be achieved by retaining more than just the (seemingly) best candidate solution if there is noise present. Nissen and Propach [59, 60] provided empirical evidence that it may be the use of a population that is distributed in search space that makes evolutionary algorithms superior to other approaches to optimization in noisy environments. However, little is known as to where the benefits of distributed populations in noisy environments stem from. According to Rechenberg [67], it would be a "little breakthrough" if a law describing the local performance of a multi-parent strategy in a quadratic fitness environment could be found.

In this chapter, we pursue a moment-based approach to analyzing the local performance of the (μ, λ)-ES in the linear environment. The linear environment is of interest as it is more simple than the spherical environment while the beneficial effects of working with distributed populations of candidate solutions can still be observed. In Section 1, we describe a moment-based approach to modeling the population of candidate solutions. In Section 2 we then set out to compute some moments of this distribution in the simple case of infinite noise strength. Finite noise strength is much more difficult to handle. In Section 3 we first show that the approach pursued for infinite noise strength does not yield results for finite noise strength. We then introduce a simplification that is equivalent to the "self-consistency conditions" imposed by Beyer [20]. While the approach does not lead to an exact solution of the problem even if an infinite number of central moments could be considered, the results that can be obtained are quite accurate provided that the kurtosis of the distribution is considered in addition to its variance and skewness. Detailed derivations of all results can be found in Appendix D in which noisy order statistics are introduced and expected moments of truncated samples are computed. The results that have been obtained are then used to arrive at quantitative insights with regard to the beneficial effects of distributed populations in the presence of noise. Finally, in Section 4, we discuss the relevance to the spherical environment of the findings that are made in the linear environment, and we numerically determine estimates for optimal population sizes and efficiencies of the (μ, λ)-ES on the noisy sphere for high search space dimensionality.

1. Modeling the Population

So as to determine the quality gain of the (μ, λ)-ES in the linear environment, the population of candidate solutions that emerges from the interplay of mutation and selection needs to be modeled appropriately. We adopt a moment-based approach similar to that introduced by Beyer [20]. Denoting the mean of the population $\{x_1, \ldots, x_\mu\}$ of candidate solutions as $\langle x \rangle = \sum_{i=1}^{\mu} x_i/\mu$, we consider the central moments

$$m_k = \frac{1}{\mu} \sum_{i=1}^{\mu} \left(x_i - \langle x \rangle \right)^k, \qquad k \geq 2, \tag{4.1}$$

of the population and study their behavior in time. The central moments of the population are random variables that characterize the population's state, and we will see that knowledge of their distribution is useful for determining the quality gain of the strategy.

Mutation and selection map the central moments of the population at time t to those at time $t + 1$. Let us first look at the effects of mutation. Mutations consist in adding independent random variables drawn from a standardized normal distribution to candidate solutions randomly selected from the population. The distribution of offspring candidate solutions is thus the convolution of a probability distribution with central moments m_k and a standardized normal distribution. According to Section 3.1 of Appendix A, the latter distribution has kth central moment $k!/(2^{k/2}(k/2)!)$ for k even and zero odd central moments. The central moments of the convolution of two probability distributions are computed in Section 2 of Appendix A. According to Equations (A.6), the first few central moments of the distribution of offspring candidate solutions are

$$\left.\begin{aligned} \mu_2 &= m_2 + 1 \\ \mu_3 &= m_3 \\ \mu_4 &= m_4 + 6m_2 + 3 \\ &\vdots \end{aligned}\right\} \tag{4.2}$$

Offspring candidate solutions are thus drawn independently from a probability distribution with central moments μ_k determined by Equations (4.2). Then, the μ seemingly best offspring candidate solutions are selected to form the population of the next time step. The influence of selection on the central moments of the population remains to be determined.

Due to the translation invariance of the linear environment, after initialization effects have faded, the distribution of the central moments of the population will approach a time-invariant limit distribution. The problem of finding that limit distribution is reminiscent of the task we faced in Section 1.1 of Chapter 3, except that now we are not looking for the stationary distribution of a scalar variable but for the stationary joint distribution of an infinite number of central moments. In practice, of course, only a finite number of moments can be considered. One can hope that the dynamics of the process are determined largely by the lower-order central moments, and that higher-order moments are of relatively little influence. In the following sections we will attempt to derive expected values of the first few central moments of the population by demanding their stationarity. We will see that in the limit of infinite noise strength, expected central moments can be determined exactly. For finite noise strength however, fluctuations of the central moments need to be ignored so as to obtain a system of equations that can be solved.

2. The Infinite Noise Limit

In the limit of infinite noise strength, there is no useful information available and therefore selection is entirely random. As all offspring candidate solutions are generated independently and from the same probability distribution, the behavior of the (μ, λ)-ES in the limit of infinite noise strength does not differ from that of the (μ, μ)-ES. While the quality gain of the (μ, μ)-ES in the linear environment is of course zero, it is nonetheless instructive to consider the central moments of the population that emerge out of the interplay of mutation and sampling from the distribution of offspring candidate solutions.

So as to obtain expected central moments of the population, let us first compute moments of the distribution of central moments of the population at time $t+1$ from central moments of the population at time t. The effect of mutations is described by Equations (4.2). It remains to be seen how picking μ candidate solutions as the parental population of the next time step modifies these values. For that purpose, sampling moments have to be computed from the moments μ_k of the distribution of offspring candidate solutions. The calculations can be found in Sections 1 and 2 of Appendix D as well as in a text by Kenney and Keeping [51]. From Equations (D.6), (D.7), (D.8), and (D.9) with Equations (4.2) it follows that

$$\mathrm{E}\left[m_2^{(t+1)}\right] = \frac{\mu-1}{\mu}\left(m_2^{(t)}+1\right) \tag{4.3}$$

$$\mathrm{E}\left[m_3^{(t+1)}\right] = \frac{(\mu-1)(\mu-2)}{\mu^2}m_3^{(t)} \tag{4.4}$$

$$\mathrm{E}\left[m_4^{(t+1)}\right] = \frac{(\mu-1)(\mu^2-3\mu+3)}{\mu^3}\left(m_4^{(t)}+6m_2^{(t)}+3\right) + \frac{3(\mu-1)(2\mu-3)}{\mu^3}\left(m_2^{(t)}+1\right)^2 \tag{4.5}$$

$$\mathrm{E}\left[m_2^{(t+1)2}\right] = \frac{(\mu-1)^2}{\mu^3}\left(m_4^{(t)}+6m_2^{(t)}+3\right) + \frac{(\mu-1)(\mu^2-2\mu+3)}{\mu^3}\left(m_2^{(t)}+1\right)^2. \tag{4.6}$$

Higher-order sampling moments could be obtained, but are irrelevant for the calculations that follow.

Let $p^{(t)}$ denote the joint probability density function of the central moments of the population at time t, and let $p_k^{(t)}$ denote the marginal probability density function of the kth central moment of the population at that time. The marginal probability density of the kth central moment of the population at time $t+1$ is

$$p_k^{(t+1)}(y) = \int p^{(t)}(\mathbf{x})p_k(y|\mathbf{x})\mathrm{d}\mathbf{x},$$

where the integration extends over all central moments considered in the analysis and where $p_k(y|\mathbf{x})$ is the conditional probability density of the kth central moment of the population at a time step given that the central moments of the population at the previous time step are $\mathbf{x} = \{m_2, m_3, m_4, \dots\}$. It follows by integration that the rth moment about zero of the kth central moment of the population at time $t+1$ is

$$\begin{aligned} \int y^r p_k^{(t+1)}(y)\mathrm{d}y &= \int y^r \int p^{(t)}(\mathbf{x})p_k(y|\mathbf{x})\mathrm{d}\mathbf{x}\mathrm{d}y \\ &= \int p^{(t)}(\mathbf{x}) \int y^r p_k(y|\mathbf{x})\mathrm{d}y\mathrm{d}\mathbf{x}. \end{aligned} \tag{4.7}$$

Expected central moments of the population can be computed from Equation (4.7) by requiring stationarity and by considering different values of k and r.

In particular, for $k = 2$ and $r = 1$, on the left hand side of Equation (4.7) we have the marginal mean $\mathrm{E}[m_2^{(t+1)}]$ of the variance of the population at time $t+1$. The value of the inner integral on the right hand side is given by Equation (4.3). It follows that

$$\begin{aligned} \mathrm{E}\left[m_2^{(t+1)}\right] &= \frac{\mu-1}{\mu} \int p^{(t)}(\mathbf{x}) \left(m_2^{(t)} + 1\right) \mathrm{d}\mathbf{x} \\ &= \frac{\mu-1}{\mu} \left(\mathrm{E}\left[m_2^{(t)}\right] + 1\right). \end{aligned} \tag{4.8}$$

Stationarity requires that $\mathrm{E}[m_2^{(t+1)}] = \mathrm{E}[m_2^{(t)}] = \mathrm{E}[m_2]$ and therefore that

$$\mathrm{E}\left[m_2\right] = \mu - 1. \tag{4.9}$$

Note that Equation (4.9) agrees with a result formerly obtained by Beyer [25].

Proceeding analogously for $k = 3$ and $r = 1$ with Equation (4.4) for the inner integral yields

$$\mathrm{E}\left[m_3^{(t+1)}\right] = \frac{(\mu-1)(\mu-2)}{\mu^2} \mathrm{E}\left[m_3^{(t)}\right]. \tag{4.10}$$

Solving for $\mathrm{E}[m_3^{(t+1)}] = \mathrm{E}[m_3^{(t)}] = \mathrm{E}[m_3]$ yields

$$\mathrm{E}[m_3] = 0. \tag{4.11}$$

This result is little surprising as a nonzero third cumulant would indicate an asymmetry not present under random selection.

Finally, for $k = 4$ and $r = 1$ one obtains with Equation (4.5) for the inner integral

$$\begin{aligned} \mathrm{E}\left[m_4^{(t+1)}\right] &= \frac{(\mu-1)(\mu^2-3\mu+3)}{\mu^3} \left(\mathrm{E}\left[m_4^{(t)}\right] + 6\mathrm{E}\left[m_2^{(t)}\right] + 3\right) \\ &+ \frac{3(\mu-1)(2\mu-3)}{\mu^3} \left(\mathrm{E}\left[m_2^{(t)2}\right] + 2\mathrm{E}\left[m_2^{(t)}\right] + 1\right). \end{aligned} \tag{4.12}$$

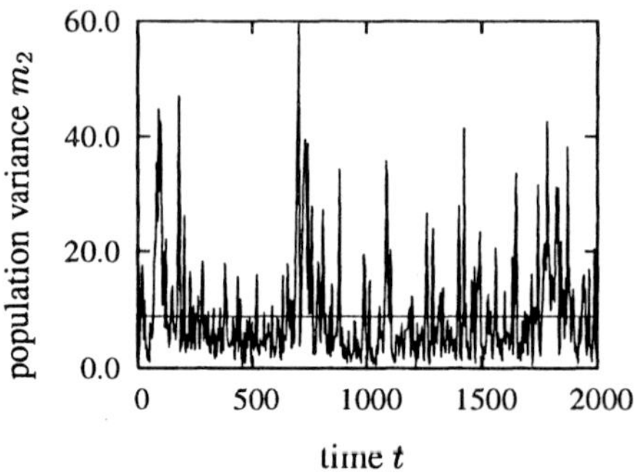

Figure 4.1. Variance m_2 of the population as a function of time t for a $(10, 10)$-ES. The rugged line has been measured empirically, the straight line reflects the expected value given by Equation (4.9).

Similarly, for $k = 2$ and $r = 2$ it follows with Equation (4.6) for the inner integral that

$$\mathrm{E}\left[m_2^{(t+1)^2}\right] = \frac{(\mu-1)^2}{\mu^3}\left(\mathrm{E}\left[m_4^{(t)}\right] + 6\mathrm{E}\left[m_2^{(t)}\right] + 3\right) + \frac{(\mu-1)(\mu^2-2\mu+3)}{\mu^3}\left(\mathrm{E}\left[m_2^{(t)^2}\right] + 2\mathrm{E}\left[m_2^{(t)}\right] + 1\right). \quad (4.13)$$

It is easily verified that demanding stationarity, i.e. requiring that $\mathrm{E}[m_4^{(t+1)}] = \mathrm{E}[m_4^{(t)}] = \mathrm{E}[m_4]$ and $\mathrm{E}[m_2^{(t+1)^2}] = \mathrm{E}[m_2^{(t)^2}] = \mathrm{E}[m_2^2]$, and solving for the unknowns using Equation (4.9) yields

$$\mathrm{E}\left[m_4\right] = \frac{3(\mu-1)^2(2\mu-1)(5\mu-6)}{6\mu^2-11\mu+6} \quad (4.14)$$

and

$$\mathrm{E}[m_2^2] = \frac{(\mu-1)(2\mu-1)(7\mu^2-11\mu+6)}{6\mu^2-11\mu+6}. \quad (4.15)$$

Considering "sample cumulants" rather than central moments, we could write alternatively

$$\mathrm{E}\left[m_4 - 3m_2^2\right] = -\frac{6\mu^2(\mu-1)(2\mu-1)}{6\mu^2-11\mu+6}. \quad (4.16)$$

In Figure 4.1, the variance m_2 of the population of a $(10, 10)$-ES is shown as a function of time. It can be seen that the distribution of the variance is rather skew and of a considerable variance, and that thus it is not very well described by its mean. The same holds true for the higher-order moments.

3. Finite Noise Strength

It is tempting to try to apply the same approach for computing expected central moments of the population to the case of finite noise strength. The first step is to compute expected central moments of the population at time $t + 1$

as functions of the central moments at time t. This task does now no longer simply involve sampling from the distribution of offspring candidate solutions. The offspring candidate solutions that are selected to form the population of the next time step are not independent. Moreover, as our analysis employs Gram-Charlier and Cornish-Fisher expansions, and as those expansions utilize standardized cumulants of the distributions involved — notably skewness and kurtosis —, the approach to be presented here relies on standardized cumulants rather than on central moments.

3.1 Determining Cumulants of the Population

As outlined in Section 2 of Appendix A, cumulants can easily be obtained from central moments. According to Equations (4.2), the offspring candidate solutions generated at time step t are independent, jointly distributed random variables with variance $m_2^{(t)} + 1$, with skewness

$$g_1^{(t)} = \frac{m_3^{(t)}}{\left(m_2^{(t)} + 1\right)^{3/2}},$$

and with kurtosis

$$g_2^{(t)} = \frac{\left(m_4^{(t)} + 6m_2^{(t)} + 3\right) - 3\left(m_2^{(t)} + 1\right)^2}{\left(m_2^{(t)} + 1\right)^2} = \frac{m_4^{(t)} - 3m_2^{(t)2}}{\left(m_2^{(t)} + 1\right)^2}.$$

Associated with every offspring candidate solution is a measured fitness value with variance ϑ^2. Making the central assumption that the probability distribution that the offspring candidate solutions are drawn from is well characterized by a few of its lower-order moments, the problem of computing expected moments of the selected offspring is tackled in Appendix D. We have chosen not include the calculations here as they are long and tedious and would divert attention from the focus of this chapter. In the appendix, noisy order statistics are introduced as a generalization of ordinary order statistics. Central moments of the selected offspring are represented as sums of products of noisy order statistics. Joint probability density functions of noisy order statistics are derived, and integral representations for the expected values of sums of products of noisy order statistics are found. Then the probability density function of the distribution that the offspring candidate solutions are drawn from is expanded in terms of derivatives of the normal distribution function, making it possible to solve the integrals in the previously obtained representation and to express the expected values of sums of products of noisy order statistics in terms of coefficients that can be obtained numerically and that depend on the population size parameters μ and λ as well as on the noise coefficient

$$a^{(t)} = \sqrt{\frac{m_2^{(t)} + 1}{m_2^{(t)} + \vartheta^2 + 1}} \tag{4.17}$$

introduced in Equation (D.18). Finally, expected moments of the selected offspring candidate solutions are obtained. From Equations (D.30), (D.31), and (D.32) it follows that

$$\mathrm{E}\left[m_2^{(t+1)}\right] = \frac{\mu - 1}{\mu}\left(m_2^{(t)} + 1\right) A_2^{(t)}, \tag{4.18}$$

that

$$\mathrm{E}\left[m_3^{(t+1)}\right] = \frac{(\mu-1)(\mu-2)}{\mu^2}\left(m_2^{(t)} + 1\right)^{3/2} A_3^{(t)}, \tag{4.19}$$

and that

$$\begin{aligned}\mathrm{E}\left[m_4^{(t+1)} - 3m_2^{(t+1)^2}\right] &= \frac{(\mu-1)(\mu-2)(\mu-3)}{\mu^3}\left(m_2^{(t)} + 1\right)^2 A_{4a}^{(t)} \\ &\quad - \frac{\mu-1}{\mu^2}\left(m_2^{(t)} + 1\right)^2 A_{4b}^{(t)},\end{aligned} \tag{4.20}$$

where the $A_j^{(t)}$ are

$$A_j^{(t)} = a_{j,0}^{(t)} + \frac{g_1^{(t)}}{6} a_{j,1}^{(t)} + \frac{g_2^{(t)}}{24} a_{j,2}^{(t)} + \ldots$$

with the coefficients $a_{i,j}$ listed in Table D.2 on page 144. All terms involving cumulants higher than the fourth as well as terms involving products of two cumulants higher than the second have been omitted and are represented by dots. Moreover, it follows from Equation (D.29) that the expected mean of the population at time $t + 1$ is

$$\mathrm{E}\left[m_1^{(t+1)}\right] = \left(m_2^{(t)} + 1\right)^{1/2} A_1^{(t)} \tag{4.21}$$

provided that the mean of the population at time t is zero.

The next step in the approach pursued in Section 2 would now require computing the expected values of the right hand sides of Equations (4.18), (4.19), and (4.20) with respect to the joint probability distribution of the variance, the skewness, and the kurtosis of the population at time t. It is obvious that these expected values do not depend on the expected variance, skewness, and kurtosis of the population at time t in a simple fashion. It is thus not possible to arrive at a system of equations that can be solved for the expected standardized cumulants.

3.2 Ignoring Fluctuations

Even though exact expected standardized cumulants of the population cannot be obtained by using the approach pursued in Section 2, it may still be worthwhile attempting to use the approach taken by Beyer [20] in the absence of noise to obtain good estimates thereof. The approach consists in neglecting fluctuations of the cumulants of the population and considering expected values only. That is, we look for values of the standardized cumulants of the population that are unchanged after mutation and selection by the operation of taking expected values, and we hope that those values do not deviate too much from the proper expected values. Demanding that

$$\mathrm{E}\left[m_2^{(t+1)}\right] = m_2^{(t)} = m_2$$

$$\mathrm{E}\left[m_3^{(t+1)}\right] = m_3^{(t)} = m_3$$

$$\mathrm{E}\left[m_4^{(t+1)} - 3m_2^{(t+1)^2}\right] = m_4^{(t)} - 3m_2^{(t)^2} = m_4 - 3m_2^2$$

and using Equations (4.18), (4.19), and (4.20) with all terms involving higher-order cumulants neglected, we have a system of three equations in three unknowns that reads

$$m_2 = \frac{\mu - 1}{\mu}(m_2 + 1)\left(a_{2,0} + \frac{g_1}{6}a_{2,1} + \frac{g_2}{24}a_{2,2}\right) \tag{4.22}$$

$$g_1 = \frac{(\mu - 1)(\mu - 2)}{\mu^2}\left(a_{3,0} + \frac{g_1}{6}a_{3,1} + \frac{g_2}{24}a_{3,2}\right) \tag{4.23}$$

$$\begin{aligned} g_2 = \frac{(\mu - 1)(\mu - 2)(\mu - 3)}{\mu^3} & \left(a_{4a,0} + \frac{g_1}{6}a_{4a,1} + \frac{g_2}{24}a_{4a,2}\right) \\ & - \frac{\mu - 1}{\mu^2}\left(a_{4b,0} + \frac{g_1}{6}a_{4b,1} + \frac{g_2}{24}a_{4b,2}\right). \end{aligned} \tag{4.24}$$

Numerically solving this system yields results for m_2, g_1, and g_2 that can be used to compute an estimate of the quality gain

$$\mathrm{E}[\Delta_{line}] \approx \sqrt{m_2 + 1}\left(a_{1,0} + \frac{g_1}{6}a_{1,1} + \frac{g_2}{24}a_{1,2}\right) \tag{4.25}$$

that follows from Equation (4.21). Note that more cumulants could be considered, but that the resulting equations would become very lengthy. As the approach does not lead to exact results no matter how many cumulants are considered, we refrain from including higher-order moments in the calculations.

In Figure 4.2, the estimates that have been obtained by solving the system of Equations (4.22), (4.23), and (4.24) and the resulting estimate for the quality

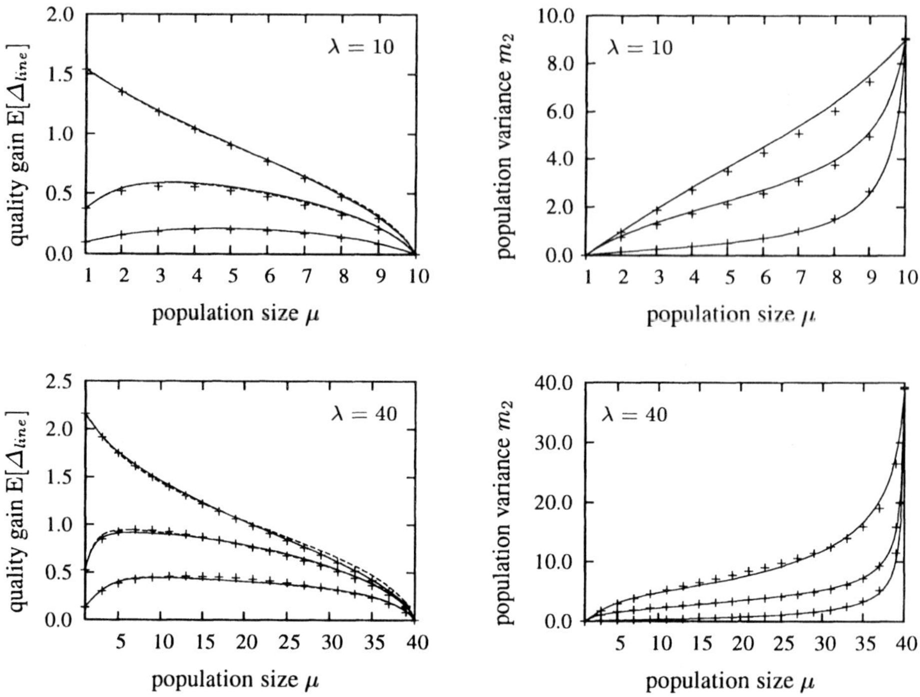

Figure 4.2. Quality gain $\mathrm{E}[\Delta_{line}]$ and variance m_2 of the population for $\lambda = 10$ (upper row) and $\lambda = 40$ (lower row) as functions of the size of the parental population μ. The lines in the right hand graphs represent the estimates obtained from solving the system of Equations (4.22), (4.23), and (4.24) for, from bottom to top, noise levels $\vartheta = 0.0$, 4.0, and 16.0. The solid lines in the graphs on the left hand side represent, from top to bottom, the corresponding estimates for $\mathrm{E}[\Delta_{line}]$ obtained from Equation (4.25). The dashed lines in the left hand graphs are the result of neglecting the terms involving skewness and kurtosis in Equation (4.25). The crosses mark data that has been obtained from empirical measurements of runs of evolution strategies in the linear environment.

gain given by Equation (4.25) are compared with measurements of runs of evolution strategies in the linear environment. It can be seen that the estimates for the quality gain and the expected variance of the population that have been obtained do agree quite closely with the values measured empirically. In fact, for $\mu = 1$, solving the system of equations yields $m_2 = g_1 = g_2 = 0$. This solution as well as the resulting value for the quality gain are exact as the parental population consists of a single point and the distribution of offspring fitness values is normal. The results agree with those found for the $(1, \lambda)$-ES by Beyer [18]. For $\vartheta = 0$, the estimates agree with those derived by Beyer [20], except that the fourth cumulant is considered here in addition. However, the results do not agree with the exact result for the expected value of the fourth cumulant that was derived in Section 2 for the special case that $\mu > 1$ and $\vartheta = \infty$.

3.3 Benefits of Distributed Populations

It can be seen from Figure 4.2 that in the presence of noise, much higher quality gain than that of the $(1, \lambda)$-ES can be achieved by choosing $\mu > 1$. As customary, we write $c_{\mu,\lambda}(\vartheta)$ for the quality gain of the (μ, λ)-ES in the linear environment. For nonzero noise, the curves in the left hand graphs have a maximum at intermediate values of μ. For example, while the quality gain of the $(1, 40)$-ES at noise level $\vartheta = 16.0$ is $c_{1,40}(16.0) \approx 0.13$, that of the $(12, 40)$-ES is $c_{12,40}(16.0) \approx 0.46$. That is, the quality gain can be increased by a factor of 3.5 without additional computational costs simply by retaining more than just the seemingly best offspring candidate solution. Rechenberg [67] empirically demonstrated that at higher noise levels even greater speed-up factors can be observed. The left hand graphs of Figure 4.2 suggest that for fixed λ, the optimal size of the parental population μ increases with increasing noise strength — a fact that was already observed by Rechenberg [67] —, and that at the same time the quality gain becomes less sensitive to the truncation ratio. Rechenberg speculated that the fact that the variance of the offspring of a (μ, λ)-ES is increased as compared with that of a $(1, \lambda)$-ES operating with the same mutation strength might contribute to the improved performance in the presence of noise of the former strategy.

So as to see what can be learned with regard to the reasons for the speed-up that can be achieved, let us consider Equation (4.25). Closer numerical investigation shows that while g_1 and g_2 are essential for obtaining a satisfactory estimate of the variance m_2 of the population, the influence of the terms they appear in Equation (4.25) is rather minor. Omitting all but the first term in the parentheses and using the definition of $a_{1,0}$ from Table D.2 and the definition of the noise coefficient from Equation (4.17) yields the approximation

$$\mathrm{E}[\Delta_{line}] = c_{\mu,\lambda}(\vartheta) \approx \frac{\sqrt{m_2 + 1}}{\sqrt{1 + \vartheta^2/(m_2 + 1)}} h^{1,0}_{\mu,\lambda}, \tag{4.26}$$

where the coefficient $h^{1,0}_{\mu,\lambda}$ agrees with the coefficient $e^{1,0}_{\mu,\lambda}$ introduced and studied by Beyer [27]. The dependence of the coefficient on μ and λ is illustrated in Figure 5.1. As can be seen from the dashed lines in the left hand graphs of Figure 4.2, the error resulting from using Equation (4.26) rather than Equation (4.25) as an estimate of the quality gain is minor.

The right hand graphs of Figure 4.2 suggest that for fixed λ, the variance of the population increases both with increasing size of the parental population μ and with increasing noise level ϑ. In the absence of noise, the quality gain according to Equation (4.26) simply reads $\mathrm{E}[\Delta_{line}] \approx \sqrt{m_2 + 1} h^{1,0}_{\mu,\lambda}$. When increasing the size of the parental population μ, the increase in $\sqrt{m_2 + 1}$ due to an increase in the variance m_2 of the population is more than offset by the decrease in $h^{1,0}_{\mu,\lambda}$ if λ remains unchanged. The corresponding curves in the left hand graphs of Figure 4.2 are decreasing monotonically. In the presence of noise, however, the variance of the population also appears under the square root in the denominator where it acts to reduce the weight of the noise-dependent term. As $m_2 + 1$ is the variance of the set of offspring candidate solutions, and as selection is based on the measured fitness of the offspring candidate solutions, the quotient $\vartheta/\sqrt{m_2 + 1}$ is the noise-to-signal ratio of the system. While for the $(1, \lambda)$-ES we have $m_2 = 0$ and the quality gain decreases with $1/\sqrt{1 + \vartheta^2}$ — a result that is known from Beyer [18] and Rechenberg [67] —, for $\mu > 1$ the noise-to-signal ratio is moderated by the nonzero variance of the population. As we have seen, the effect of this decrease of the noise-to-signal ratio can outweigh the decrease in $\sqrt{m_2 + 1}h_{\mu,\lambda}^{1,0}$ that results from an increase of μ. Equation (4.26) thus not only provides a quantitative background for Rechenberg's speculation, but it also demonstrates that the increased variance of the offspring is the *only* reason for the improved performance of the (μ, λ)-ES as compared with the $(1, \lambda)$-ES.

4. The Spherical Environment

As noted in Section 3 of Chapter 2, the linear environment is of little interest for efficiency considerations. The efficiency of evolution strategies in the linear environment is proportional to the mutation strength and can therefore be increased indefinitely. For the spherical environment, we have seen in Section 5 of Chapter 2 that in the limit of infinite search space dimensionality the fitness advantage of an offspring candidate solution equals the sum of a normally distributed term that stems from the central component of the mutation that generates the candidate solution and a negative constant term that can be traced back to the mutation's lateral component. That latter term limits the mutation strengths that evolution strategies can achieve a positive quality gain with. In the present section, we give a quality gain law for the (μ, λ)-ES on the noisy sphere by reducing the problem to a single dimension and using results from the preceding section. While a number of simplifications need to be made in its derivation, empirical measurements suggest that the accuracy of the law is very good provided that the dimensionality of the search space is sufficiently high.

4.1 Determining the Quality Gain

The determination of the quality gain as defined in Section 4 of Chapter 2 is complicated considerably by the fact that it involves computing the centroid of the population of candidate solutions and thus requires modeling the distribution of the population in the N-dimensional search space. So as to reduce the problem to a one-dimensional one, in this section we average fitness values rather than candidate solutions. That is, instead of defining the quality gain as the expected change in fitness of the centroid, we define it as the expected change in average fitness of the population. In that way, there are no interactions in search space between the candidate solutions comprising a population and all calculations can be performed in the (one-dimensional) space of fitness values. While this makes it impossible to directly compare the quality gain of the (μ, λ)-ES with that of the other strategies analyzed in this work, empirical evidence suggests that the difference between the two definitions becomes more and more irrelevant as the search space dimensionality increases.

Let the population at time step t consist of candidate solutions with fitness values R_i^2, $i = 1, \dots, \mu$, and let $R^2 = \sum_{i=1}^{\mu} R_i^2/\mu$ denote the average fitness value. Offspring candidate solutions with fitness values r_j^2, $j = 1, \dots, \lambda$, are generated by randomly choosing a parent and adding a mutation vector with mutation strength σ. Using the decomposition of mutation vectors outlined in Section 5 of Chapter 2, it follows in analogy to Equation (2.11) that the fitness of an offspring generated from parent i is

$$r^2 = (R_i - \sigma z_A)^2 + \sigma^2 \|\mathbf{z}_B\|^2,$$

where z_A is the signed length of the central component of the mutation vector used to generate the offspring candidate solution and where $\|\mathbf{z}_B\|^2$ is the squared length of its lateral component. Assuming large N and making the same simplifications as in Section 5 of Chapter 2, it follows in analogy to Equation (2.14) that in the limit of infinite search space dimensionality the normalized fitness advantage associated with the offspring candidate solution is

$$\begin{aligned} \delta^* &= N\frac{R_i^2 - r^2}{2R^2} \\ &= \sigma^* \frac{R_i}{R} z_A - \frac{\sigma^{*2}}{2}, \end{aligned}$$

where the normalized mutation strength is $\sigma^* = \sigma N/R$. Mutations thus result in a negative term $-\sigma^{*2}/2$ that is the same no matter what the parent of an offspring candidate solution is, and a term drawn from a normal distribution with mean zero. The negative term is without influence on the central moments of the distribution of the distances R_i of the candidate solutions from the optimizer. We could apply our results from Section 3 if the variance of the normally distributed term were the same for all offspring. Unfortunately, that variance is $(\sigma^* R_i/R)^2$ and thus depends on the parental candidate solution that the offspring is generated from. However, it can be observed that the variance of the fitness values of the population decreases with increasing search space dimensionality. This is plausible as we have seen in Sections 2 and 3 that in the linear environment the variance of the population of candidate solutions that results from the interplay of mutation and selection is bounded from above by

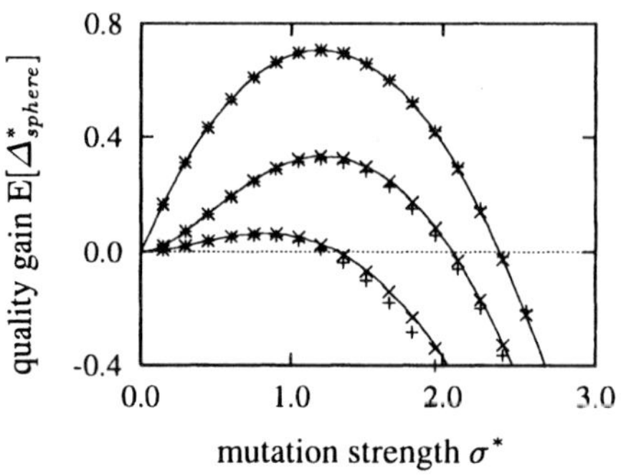

Figure 4.3. Normalized quality gain E[Δ^*_{sphere}] of a $(3, 10)$-ES in the spherical environment as a function of normalized mutation strength σ^* for, from top to bottom, normalized noise strengths $\sigma^*_\epsilon = 0.0, 2.0$, and 4.0. The solid lines mark results from Equation (4.27) with the coefficients $c_{\mu,\lambda}(\vartheta)$ measured empirically in the linear environment. The crosses represent data from runs of evolution strategies with $N = 40$ $(+)$ and $N = 400$ $(\times)$.

the size of the parental population. Assuming that the same is true for the R-values in the spherical environment and taking into account that in the spherical environment the mutation strength is σ rather than unity, it seems reasonable to assume that the quotient $d_i = (R_i - R)/\sigma$ is of bounded variance. As

$$\frac{R_i}{R} = 1 + \frac{\sigma}{R} d_i = 1 + \frac{\sigma^*}{N} d_i$$

it follows that R_i/R more and more closely approaches unity as the search space dimensionality increases. Thus the quality gain of the (μ, λ)-ES on the noisy sphere in the limit of infinite search space dimensionality is well approximated by

$$\mathrm{E}[\Delta^*_{sphere}] = \sigma^* c_{\mu,\lambda}(\vartheta) - \frac{\sigma^{*2}}{2}, \tag{4.27}$$

where $\vartheta = \sigma^*_\epsilon/\sigma^*$ and where $c_{\mu,\lambda}(\vartheta)$ denotes the quality gain of the (μ, λ)-ES with unity mutation strength in the linear environment. For zero noise strength, Equation (4.27) formally agrees with a progress rate law given by both Rechenberg [67] and Beyer [27]. Rechenberg also claimed the validity of the law in the presence of noise.

Figure 4.3 compares predictions from Equation (4.27) with empirical measurements of the quality gain of a $(3, 10)$-ES in the spherical environment with search space dimensionalities $N = 40$ and $N = 400$. As the focus here is the relevance of the linear quality gain for the spherical environment, empirical values for the coefficients $c_{\mu,\lambda}(\vartheta)$ have been used so as to not introduce errors resulting from the approximation made in Section 3. It can be seen that for $N = 40$ the agreement is quite good, but that the accuracy of the predictions afforded by Equation (4.27) decreases with increasing noise strength and with increasing mutation strength. Search space dimensionality $N = 400$ is sufficiently high for achieving very good agreement with empirical measurements across the entire range of noise and mutation strengths considered.

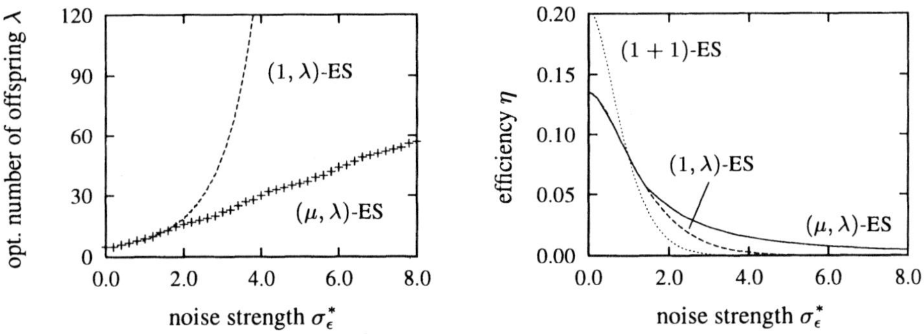

Figure 4.4. Optimal number of offspring per time step λ and maximal efficiency η on a high-dimensional sphere as functions of normalized noise strength σ_ϵ^*. In the left hand graph, the dashed line represents results for the $(1, \lambda)$-ES, the crosses for the (μ, λ)-ES with optimally chosen μ. In the right hand graph, the solid and dashed lines correspond to the (μ, λ)-ES and the $(1, \lambda)$-ES, respectively, with optimally chosen population size parameters. The dotted line reflects the result for the $(1 + 1)$-ES that was obtained in Chapter 3.

4.2 Optimizing the Efficiency

Let us now determine optimal population sizes and maximal efficiencies in the spherical environment for sufficiently high search space dimensionality. For that purpose, we determine numerically mutation strengths and population size parameters μ and λ that maximize the efficiency

$$\eta = \frac{1}{\lambda} \mathrm{E}[\Delta^*_{sphere}]$$

of the (μ, λ)-ES, where the quality gain is taken to be determined by Equation (4.27). So as to not introduce errors resulting from inaccurate quality gain predictions, we determine the coefficients $c_{\mu,\lambda}(\vartheta)$ from empirical measurements in the linear environment rather than by using the approach from Section 3. While for low normalized noise strengths $(0.0 \leq \sigma_\epsilon^* \leq 4.0)$ there are virtually no discrepancies between the resulting curves, for higher normalized noise strengths the dependence of the quality gain on the population size parameters becomes so weak that some deviations in the results can be observed.

The left hand graph of Figure 4.4 shows the optimal number of offspring λ as a function of the normalized noise strength for the (μ, λ)-ES with optimally chosen μ as well as for the $(1, \lambda)$-ES. It can be seen that, except for very low normalized noise strengths, the (μ, λ)-ES ideally operates with many fewer offspring candidate solutions per time step than the $(1, \lambda)$-ES. At least for the range considered, the relationship between the normalized noise strength and the optimal number of offspring candidate solutions of the (μ, λ)-ES appears to be nearly linear. The optimal truncation ratio ranges between one twelfth and a quarter.

The right hand graph of Figure 4.4 compares the efficiency of the (μ, λ)-ES with optimally chosen population size parameters μ and λ with those of the $(1, \lambda)$-ES with optimally chosen λ and of the $(1+1)$-ES. The efficiency of the $(1+1)$-ES exceeds the efficiency of the other two strategies only up to a normalized noise strength of $\sigma_\epsilon^* \approx 1.0$ and is markedly inferior for higher noise strengths. Up to a normalized noise strength of about $\sigma_\epsilon^* \approx 1.4$ it is not useful to retain more than a single candidate solution and the curves for the (μ, λ)-ES and the $(1, \lambda)$-ES agree. Above this noise strength the efficiency of the (μ, λ)-ES can significantly exceed that of the $(1, \lambda)$-ES.

Chapter 5

THE $(\mu/\mu, \lambda)$-ES: GENETIC REPAIR

An important component of evolution strategies that has been left unconsidered in the analyses presented so far is recombination. In the present chapter, we study the effects of global intermediate recombination by investigating the local performance of the $(\mu/\mu, \lambda)$-ES in the spherical environment. Considering this particular form of recombination greatly simplifies the analysis as it involves computing the centroid of the entire parental population as the starting point that all mutations originate at. Thus, there is no need to model distributed populations of candidate solutions.

In the absence of noise, the effects of global intermediate recombination in the spherical environment have been studied by Rechenberg [67] as well as by Beyer [21]. Their findings can be summarized briefly as follows: global intermediate recombination averages the mutation vectors that belong to those candidate solutions that are selected to form the population of the next time step. While the central components of those mutation vectors are correlated, their lateral components are not. Averaging leads to a progress vector with a lateral component the length of which is reduced substantially compared with the lengths of the lateral components of the mutation vectors. The amount of this reduction in length depends on the size of the population. As the lateral component of the progress vector contributes a negative term to the quality gain of the $(\mu/\mu, \lambda)$-ES, the process of averaging is beneficial to the local performance of the strategy. Beyer has coined the term *genetic repair principle* for that effect. As a consequence of the presence of genetic repair, the strategy can operate with mutation strengths that exceed those useful without recombination and achieve considerably improved performance. An additional result obtained by Beyer [27] in the absence of noise is that in the spherical environment for sufficiently high search space dimensionality, the efficiency of the $(\mu/\mu, \lambda)$-ES cannot exceed that of the $(1 + 1)$-ES. The latter efficiency is only approached asymptotically as the population size increases. Beyer [21] has also shown that in search spaces of finite dimensionality, predictions resulting from the assumption of infinite dimensionality can severely overestimate the true performance of the strategy, and he presented an improved progress rate law that better accounts for finite search space dimensionalities.

In this chapter, we extend Beyer's analysis to include nonzero noise and cumulative mutation strength adaptation. In Section 1, we study the performance of the $(\mu/\mu, \lambda)$-ES using the simplifications that hold for the spherical environment in the limit of infinite search space dimensionality and that were introduced in Section 5 of Chapter 2. A concise quality gain formula that neatly generalizes previous findings and that reveals interesting scaling properties of the $(\mu/\mu, \lambda)$-ES is derived. However, this quality gain formula is rather inaccurate for search spaces of moderate dimensionality, and its value for all but the smallest populations is limited. Therefore, in Section 2, we extend the simple model from Section 5 of Chapter 2 to include some of the terms that were left unaccounted for previously. We then derive a more accurate quality gain formula and study its implications numerically. Finally, in Section 3 we investigate the performance and the effects that noise has on it of the cumulative mutation strength adaptation scheme by Hansen and Ostermeier [44].

1. Simple Performance Analysis

In this section, we determine the quality gain of the $(\mu/\mu, \lambda)$-ES in the spherical environment using the simple model assuming infinite search space dimensionality that was introduced in Section 5 of Chapter 2. We then discuss implications of the results of that analysis. Finally, we address the tradeoff between explicit averaging over multiple fitness measurements so as to reduce noise and working with a larger population size.

1.1 Determining the Quality Gain

From Equation (2.10) it can be seen that the one-generation gain

$$\Delta_{sphere} = f(\langle \mathbf{x} \rangle) - f(\langle \mathbf{x} \rangle + \sigma \langle \mathbf{z} \rangle)$$

of the $(\mu/\mu, \lambda)$-ES that was introduced in Section 3 of Chapter 2 is the fitness advantage associated with the progress vector $\langle \mathbf{z} \rangle$. Thus, the quality gain of the $(\mu/\mu, \lambda)$-ES in the spherical environment is the expected fitness advantage associated with that vector. The progress vector equals the average of those mutation vectors that correspond to offspring candidate solutions that are selected for survival. As mutation vectors, progress vectors can be decomposed as illustrated in Figure 2.2. The central and lateral components of the progress vector are the averages of the respective components of the mutation vectors being averaged. Writing $\langle z_A \rangle$ for the signed length of the central component of the progress vector and using the normalizations introduced in Equations (2.8)

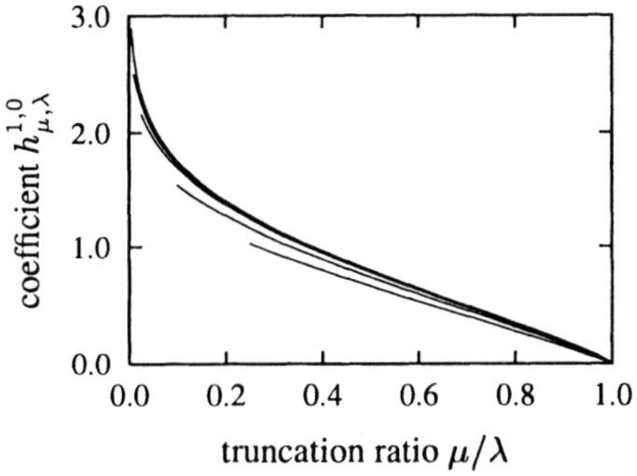

Figure 5.1. Coefficients $h^{1,0}_{\mu,\lambda}$ defined in Equation (D.27) as functions of the truncation ratio μ/λ for different values of λ. The curves correspond to, from bottom to top, $\lambda = 4, 10, 40, 100$, and the limit case $\lambda = \infty$ described by Equation (5.5). The curves are displayed in the range extending from $1/\lambda$ to 1.0. Note that only values μ/λ with integer μ are of interest here.

and (2.12), it follows from Equation (2.13) that the normalized one-generation gain is

$$\Delta^*_{sphere} = \sigma^* \langle z_A \rangle - \frac{\sigma^{*2}}{2N} \| \langle \mathbf{z} \rangle \|^2. \tag{5.1}$$

To determine the quality gain of the strategy, it remains to compute the expected signed length of the central component as well as the expected overall squared length of the progress vector.

Recall from Section 5 of Chapter 2 that in the limit of infinite search space dimensionality, the lateral components of all mutation vectors are of equal length and of random direction in the $(N-1)$-dimensional hyperplane defined by normal vector $\hat{\mathbf{x}} - \langle \mathbf{x} \rangle$. Moreover, we had seen that the contribution of the central component of a mutation vector to that vector's overall squared length is negligible provided that the search space dimensionality is sufficiently high. The contribution of the overall squared length of a mutation vector to the fitness advantage associated with that vector is a constant and therefore without influence on the selection mechanism. Selection is influenced by the central components of mutation vectors and by noise only.

The squared length of the progress vector is the sum of the squared lengths of that vector's central and lateral components. As the central and lateral components of the progress vector are the averages of the respective components of the mutation vectors being averaged, the considerations made for mutation vectors in Section 5 of Chapter 2 hold for progress vectors as well. In particular, the contribution of the central component to the overall squared length of a progress vector is negligible provided that the search space dimensionality is sufficiently high. Averaging μ independent random vectors of squared length N yields a random vector of squared length N/μ. Thus, the expected squared length of the progress vector is

$$\mathrm{E}\left[\| \langle \mathbf{z} \rangle \|^2 \right] = \frac{N}{\mu}. \tag{5.2}$$

The central components of the mutation vectors are drawn independently from a standardized normal distribution. There is noise of relative strength $\vartheta = \sigma^*_\epsilon / \sigma^*$

interfering in the selection process as selection of candidate solutions based on their normalized fitness advantage involves a signal of strength σ^* resulting from the central component of the corresponding mutation vector and noise of strength σ_ϵ^*. The mean of those central components that are selected based on their measured fitness advantage is to be computed. We thus have the situation described in Section 5 of Appendix D. From Equation (D.33) it follows that

$$\mathrm{E}[\langle z_A \rangle] = \frac{h_{\mu,\lambda}^{1,0}}{\sqrt{1+\vartheta^2}}. \tag{5.3}$$

Thus, according to Equation (5.1), the quality gain of the $(\mu/\mu, \lambda)$-ES in the spherical environment in the limit of infinite search space dimensionality is

$$\mathrm{E}[\Delta_{sphere}^*] = \sigma^* c_{\mu/\mu,\lambda}(\vartheta) - \frac{\sigma^{*2}}{2\mu}, \tag{5.4}$$

where

$$c_{\mu/\mu,\lambda}(\vartheta) = \frac{h_{\mu,\lambda}^{1,0}}{\sqrt{1+\vartheta^2}}.$$

Note the presence of the factor μ in the denominator of the second term on the right hand side of Equation (5.4) that indicates the presence of genetic repair. The coefficients $h_{\mu,\lambda}^{1,0}$ agree with the coefficients $c_{\mu/\mu,\lambda}$ that have been studied by Beyer [27]. In particular, Beyer has shown that for large values of λ, the approximation

$$h_{\mu,\lambda}^{1,0} \approx \frac{\lambda}{\sqrt{2\pi}\mu} \exp\left(-\frac{1}{2}\left[\Phi^{-1}\left(1-\frac{\mu}{\lambda}\right)\right]^2\right), \tag{5.5}$$

where Φ^{-1} is the inverse of the distribution function of the standardized normal distribution, holds. The dependence of the coefficients on the truncation ratio μ/λ is illustrated in Figure 5.1. It can be seen that Equation (5.5) provides a reasonably exact estimate for $h_{\mu,\lambda}^{1,0}$ even for moderate values of λ.

1.2 Discussion

Equation (5.4) generalizes several previously obtained results and confirms a number of previously made hypotheses and observations. In particular, for $\sigma_\epsilon^* = 0$, Equation (5.4) agrees formally with the result of the progress rate analysis of the $(\mu/\mu, \lambda)$-ES on the noise-free sphere published by Rechenberg [67] and by Beyer [21]. For $\mu = 1$, results of the analysis of the $(1, \lambda)$-ES on the noisy sphere that have been derived by Beyer [18] and by Rechenberg [67] and that also appeared as a special case in Chapter 4 appear as special cases. Moreover, solving Equation (5.4) for the distance R to the optimizer proves the validity of a hypothesis regarding the residual location error made by Beyer [26]. Finally,

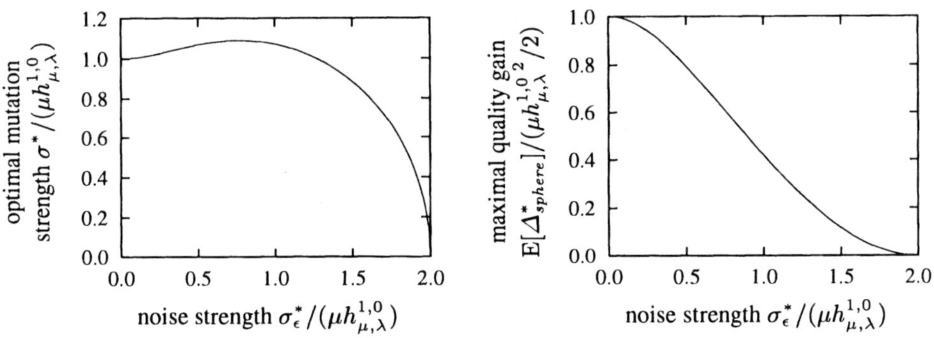

Figure 5.2. Optimal normalized mutation strength σ^* and maximal normalized quality gain $\mathrm{E}[\Delta^*_{sphere}]$ as functions of normalized noise strength σ^*_ϵ. Note the scaling of the axes: normalized mutation strength and normalized quality gain are scaled by their respective optimal values in the absence of noise; the normalized noise strength is scaled by the optimal normalized mutation strength in the absence of noise.

Equation (5.4) confirms empirical findings regarding the performance of the $(\mu/\mu, \lambda)$-ES made by Herdy that are quoted by Rechenberg [67].

Figure 5.2 has been obtained numerically from Equation (5.4) and shows the optimal normalized mutation strength σ^* and the maximal normalized quality gain $\mathrm{E}[\Delta^*_{sphere}]$ as functions of the normalized noise strength σ^*_ϵ. For the $(1, \lambda)$-ES the graphs have already been shown by Rechenberg [67]. Note that the axes have been scaled so as to make the graphs independent of the population size parameters. Only in the special case of zero noise strength can the optimal values

$$\sigma^* = \mu h^{1,0}_{\mu,\lambda} \qquad \text{and} \qquad \mathrm{E}[\Delta^*_{sphere}] = \frac{\mu h^{1,0\,2}_{\mu,\lambda}}{2}$$

be obtained analytically. It can be seen that a $(\mu/\mu, \lambda)$-ES is capable of positive quality gain up to a normalized noise strength of $2\mu h^{1,0}_{\mu,\lambda}$. Considering the fact that according to Figure 5.1 the coefficient $h^{1,0}_{\mu,\lambda}$ depends only very weakly on the population size if the truncation ratio is fixed, the normalized noise strength up to which positive quality gain is possible is thus roughly proportional to μ. Therefore, provided that the search space dimensionality is sufficiently high for the simplifications made in the derivation of Equation (5.4) to hold, positive quality gain can be achieved for any noise strength by increasing the population size.

Figure 5.3 shows both the optimal truncation ratio as a function of the normalized noise strength σ^*_ϵ and the efficiency η as a function of the number of offspring λ. It, too, has been obtained numerically from Equation (5.4). For the left hand graph, the approximation for the coefficients $h^{1,0}_{\mu,\lambda}$ that is given by Equation (5.5) has been used. It can be seen that the optimal truncation ratio increases with increasing noise strength from its value of about 0.270 in the

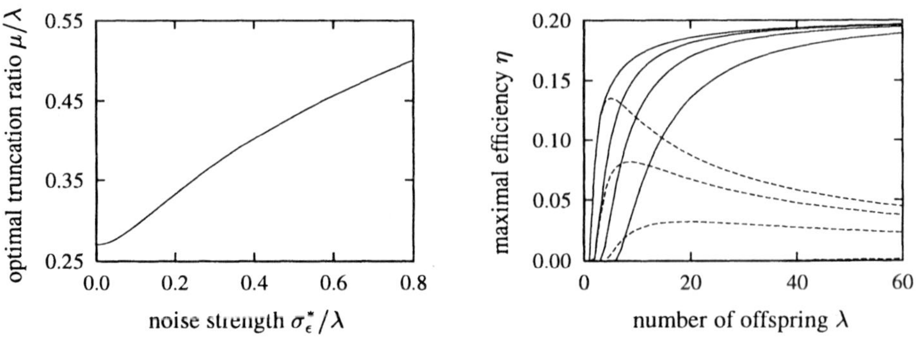

Figure 5.3. Optimal truncation ratio μ/λ as a function of normalized noise strength σ_ϵ^* and maximal efficiency η as a function of the number of offspring per time step λ. Note the scaling of the abscissa in the left hand graph. The lines in the right hand graph correspond to, from top to bottom, normalized noise strengths $\sigma_\epsilon^* = 0.0$, 1.0, 2.0, and 4.0. The solid lines represent results for the $(\mu/\mu, \lambda)$-ES with optimally tuned truncation ratio, the dashed lines represent the corresponding results for the $(1, \lambda)$-ES.

absence of noise to 0.5 at the point where no positive quality gain is possible anymore. The $(\mu/\mu, \lambda)$-ES can thus partially compensate for a lack of reliable information by using the noisy information provided by a larger number of parents than it would optimally use in the absence of noise. The right hand graph shows that by sufficiently increasing the number of offspring generated per time step (and along with it the population size μ), the efficiency of the $(\mu/\mu, \lambda)$-ES can be increased up to its maximal value of about 0.202 in the absence of noise. This is in marked contrast to the $(1, \lambda)$-ES the efficiency of which is indicated by dashed lines and that starts to decline above a certain value of λ. It has to be noted however that the simplifications made in the derivation of Equation (5.4) are rendered invalid by too large a population size. The significance of the result in finite-dimensional search spaces is therefore limited. We will need to readdress the issue of comparing optimal efficiencies in Section 2 where we expand the simple model from Section 5 of Chapter 2 to include terms that improve its accuracy for finite N.

The reason for the improved performance of the $(\mu/\mu, \lambda)$-ES as compared to the $(1, \lambda)$-ES is — as in the absence of noise — the genetic repair effect. The factor μ in the denominator of the loss term in Equation (5.4) reflects the reduction of the length of the lateral components of the mutation vectors brought about by recombination. As a consequence, it is possible to explore the search space at higher mutation strengths. In the absence of noise, as has been seen above, the increase in optimal mutation strength is about μ-fold. In a noisy environment, the increased mutation strengths have the additional benefit of leading to a reduced noise-to-signal ratio $\vartheta = \sigma_\epsilon^*/\sigma^*$. In Chapter 4 we have seen that the (μ, λ)-ES achieves increased signal strengths by virtue of the population of candidate solutions being distributed in search space; the increased signal

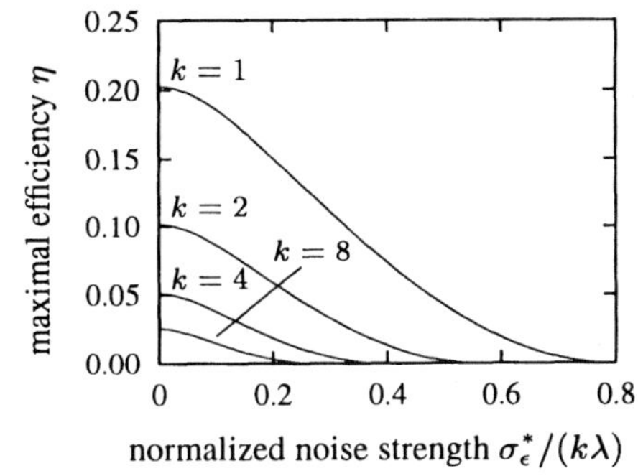

Figure 5.4. Maximal efficiency η as a function of normalized noise strength σ_ϵ^* for several $(\mu/\mu, \lambda)$-ES that average over k independent fitness measurements. Note the scaling of the abscissa. The curves display, from top to bottom, the results for strategies with k-times averaging and λ/k offspring candidate solutions per time step for $k = 1, 2, 4$, and 8. The population size μ is optimally tuned.

strength that makes the $(\mu/\mu, \lambda)$-ES effective in the presence of noise on the other hand stems from the possibility of using higher mutation strengths due to genetic repair. The advantages of increased mutation strengths in noisy environments had been recognized before and had led to the formulation of $(1, \lambda)$-ES using *rescaled mutations* suggested by Rechenberg [67] and analyzed by Beyer [24, 26]. The present analysis has shown that the positive effect of an increased mutation strength can be achieved naturally when using $(\mu/\mu, \lambda)$-ES, and that no rescaling is required.

1.3 Resampling and Population Sizing

In Section 3.2 of Chapter 3, we have discussed the question of whether resampling, i.e. averaging over a number of independent fitness measurements, is a useful way of coping with noise. For the $(1+1)$-ES, we found that above a certain noise strength resampling is indeed beneficial to the quality gain. Similarly, it is possible to ask whether resampling is preferable to increasing the population size of a $(\mu/\mu, \lambda)$-ES in the presence of noise. For the $(1, \lambda)$-ES, Beyer [18] found a generally affirmative answer to this question. Followup research by Hammel and Bäck [41] not only found empirical support for Beyer's results, but they also reported that better performance could be obtained by averaging over multiple samples rather than by increasing the population size for a multi-parent strategy with self-adaptive mutation strength that employs a nontrivial form of recombination. On the basis of these results, it is hard to justify considering an evolution strategy an algorithm which works well in noisy environments. After all, removing the noise seems to be more efficient than letting the evolution strategy deal with it.

We can compare the effects of resampling and of increasing the population size on the maximal efficiencies that can be achieved by considering Equation (5.4). The quality gain of a $(\mu/\mu, \lambda)$-ES that uses k-times resampling is given by Equation (5.4) with normalized noise strength $\sigma_\epsilon^*/\sqrt{k}$. Figure 5.4 compares the efficiencies of different $(\mu/\mu, \lambda)$-ES that use k-times resampling and that generate λ/k offspring candidate solutions per time step for different values of k. The approximation for the coefficients $h_{\mu,\lambda}^{1,0}$ that is given by Equation (5.5) has been used. Note that the number of fitness function evaluations per time step is the same for all strategies. It can be seen from the figure that the strategy that relies on a large population size rather than on a reduction of the noise strength by means of averaging consistently outperforms strategies with $k > 1$. According to Equation (5.4), the $(\mu/\mu, \lambda)$-ES achieves positive quality gain if

$$\sigma^{*2} + \sigma_\epsilon^{*2} < \left(2\mu h_{\mu,\lambda}^{1,0}\right)^2 .$$

Therefore, it is a necessary prerequisite that

$$\sigma_\epsilon^* < 2\mu h_{\mu,\lambda}^{1,0}$$

holds. Thus, a $(\mu/\mu, \lambda)$-ES that averages values obtained in k fitness function evaluations and that operates at normalized noise strength σ_ϵ^* is capable of operating with a positive efficiency if $2\sqrt{k}\mu h_{\mu,\lambda}^{1,0} > \sigma_\epsilon^*$. A $(k\mu/k\mu, k\lambda)$-ES operating in the same environment that uses fitness values without averaging is capable of operating with a positive efficiency if $2k\mu h_{k\mu,k\lambda}^{1,0} > \sigma_\epsilon^*$. As for sufficiently large λ the coefficients $h_{\mu,\lambda}^{1,0}$ and $h_{k\mu,k\lambda}^{1,0}$ approximately agree, and as for $k > 1$ the relation $k > \sqrt{k}$ holds, it is preferable to increase μ and λ rather than to average over several measured fitness values. This is an encouraging result that shows that the $(\mu/\mu, \lambda)$-ES indeed has the potential to deal with noise in an effective manner. The observations reported by Hammel and Bäck [41] are likely to be attributable to the insufficient performance of the mutation strength adaptation scheme they employed.

2. Improving the Accuracy

While the simple model introduced in Section 5 of Chapter 2 may be appropriate for studying the local performance in the spherical environment of the $(1+1)$-ES and of the (μ, λ)-ES, the accuracy of Equation (5.4) is often insufficient for describing the local performance of the $(\mu/\mu, \lambda)$-ES. Large mutation strengths render some of the assumptions made in Section 5 of Chapter 2 invalid. Only for very small populations and for very high search space dimensionalities do the results obtained in the limit of infinite search space dimensionality properly reflect features of the strategy in finite-dimensional search spaces. In the present section, we attempt to alleviate some of the deficiencies of the simple model by removing or altering some of the simplifications made above. We will see that the predictions that result are much more accurate. The improved quality gain law lacks the conciseness and elegance of that derived in the previous section, but it nonetheless has important implications that can be studied numerically.

2.1 Determining the Quality Gain

In Section 5 of Chapter 2, we have made two simplifications that are valid in the limit of infinite search space dimensionality, but that introduce inaccuracies for finite N. First, we ignored the fluctuations of the squared length of the lateral component of mutation vectors, and second, we neglected the influence of the contribution of the central component to the overall squared length of mutation vectors. Comparatively large inaccuracies for the $(\mu/\mu, \lambda)$-ES arise both from the facts that the length of the lateral component of the progress vector is reduced by genetic repair, and that the $(\mu/\mu, \lambda)$-ES operates at higher mutation strengths than those strategies that do not employ recombination.

As a step towards improving the accuracy, let us model the squared length of the lateral component of mutation vectors by a normal variate with mean N and with variance $2N$ rather than simply by N. That is, we now take fluctuations of the length of the lateral component of mutation vectors into account. According to Section 3.2 of Appendix A, this level of approximation corresponds to considering the first term in the Gram-Charlier expansion of the χ_N^2-distribution that the squared length is drawn from. Still ignoring the contribution of the central component to the squared length, the normalized measured fitness advantage associated with a mutation vector is

$$\delta_\epsilon^* = \sigma^* z_A - \frac{\sigma^{*2}}{2} + \frac{\sigma^{*2}}{\sqrt{2N}} z_B + \sigma_\epsilon^* z_\epsilon, \tag{5.6}$$

where z_A, $z_B = (N - \|\mathbf{z}_B\|^2)/\sqrt{2N}$, and z_ϵ are drawn independently from a standardized normal distribution. The term involving z_B models fluctuations of the squared length of the lateral component, the term involving z_ϵ is a result of noise. To compute the expected signed length of the central component of the progress vector, let us write Equation (5.6) as

$$\delta_\epsilon^* = \sigma^* \left(z_A + \frac{\sigma^*}{\sqrt{2N}} z_B + \vartheta z_\epsilon \right) - \frac{\sigma^{*2}}{2},$$

where as before $\vartheta = \sigma_\epsilon^*/\sigma^*$. As the selection mechanism is indifferent to the linear transformation, and as $\sigma^* z_B/\sqrt{2N} + \vartheta z_\epsilon$ is normally distributed with mean zero and with variance $\sigma^{*2}/2N + \vartheta^2$, we again have the situation described in Section 5 of Appendix D. From Equation (D.33) it follows that the expected average of the central components of the selected offspring and therefore the expected signed length of the central component of the progress vector is

$$\mathrm{E}\left[\langle z_A \rangle\right] = \frac{h_{\mu,\lambda}^{1,0}}{\sqrt{1 + \vartheta^2 + \sigma^{*2}/2N}}. \tag{5.7}$$

Comparison with Equation (5.3) shows that taking the variance of the squared length of the lateral component of mutation vectors into account reduces the estimate for the expected signed length of the central component of the progress vector as witnessed by the presence of the third summand under the square root. Fluctuations of the length of the lateral component of mutation vectors act as additional noise interfering in the selection process.

The lateral component of the progress vector is the average of the lateral components of the mutation vectors corresponding to the selected offspring candidate solutions. Writing $\mathbf{z}^{(k;\lambda)}$, $k = 1, \ldots, \mu$, for those mutation vectors that correspond to offspring candidate solutions that are selected, the expected squared length of the lateral component of the progress vector is

$$\begin{aligned} \mathrm{E}\left[\|\langle \mathbf{z}_B \rangle\|^2\right] &= \frac{1}{\mu^2} \sum_{i=2}^{N} \mathrm{E}\left[\left(\sum_{k=1}^{\mu} z_i^{(k;\lambda)}\right)^2\right] \\ &= \frac{1}{\mu^2} \sum_{i=2}^{N} \sum_{k=1}^{\mu} \mathrm{E}\left[\left(z_i^{(k;\lambda)}\right)^2\right] + \frac{1}{\mu^2} \sum_{i=2}^{N} \sum_{j \neq k} \mathrm{E}\left[z_i^{(j;\lambda)} z_i^{(k;\lambda)}\right]. \end{aligned}$$

As the lateral components of the selected offspring candidate solutions are independent, all summands in the last double sum equal zero. Exchanging the order of the summations of the other double sum yields

$$\mathrm{E}\left[\|\langle \mathbf{z}_B \rangle\|^2\right] = \frac{1}{\mu^2} \sum_{k=1}^{\mu} \mathrm{E}\left[\|\mathbf{z}_B^{(k;\lambda)}\|^2\right].$$

Writing $\langle z_B \rangle = \sum_{k=1}^{\mu} z_B^{(k;\lambda)}/\mu$ and making use of $\|\mathbf{z}_B\|^2 = N - \sqrt{2N} z_B$ shows that

$$\mathrm{E}\left[\|\langle \mathbf{z}_B \rangle\|^2\right] = \frac{1}{\mu}\left(N - \sqrt{2N}\mathrm{E}[\langle z_B \rangle]\right). \tag{5.8}$$

It remains the task of determining $\mathrm{E}[\langle z_B \rangle]$. In analogy to the procedure for computing the expected signed length of the central component of the progress vector, let us write Equation (5.6) as

$$\delta_\epsilon^* = \frac{\sigma^{*2}}{\sqrt{2N}}\left(z_B + \frac{\sqrt{2N}}{\sigma^*}(z_A + \vartheta z_\epsilon)\right) - \frac{\sigma^{*2}}{2}.$$

As $\sqrt{2N}(z_A + \vartheta z_\epsilon)/\sigma^*$ is normally distributed with mean zero and with variance $2N(1 + \vartheta^2)/\sigma^{*2}$, it follows in analogy to Equation (5.7) from Equation (D.33) that the expected average of the z_B values of the selected offspring is

$$\mathrm{E}\left[\langle z_B \rangle\right] = \frac{h_{\mu,\lambda}^{1,0}}{\sqrt{1 + 2N/\sigma^{*2} + 2N\vartheta^2/\sigma^{*2}}}.$$

Thus, it follows from Equation (5.8) that the expected squared length of the lateral component of the progress vector is

$$\mathrm{E}\left[\|\langle \mathbf{z}_B\rangle\|^2\right] = \frac{1}{\mu}\left(N - \frac{\sigma^* h^{1,0}_{\mu,\lambda}}{\sqrt{1+\vartheta^2+\sigma^{*2}/2N}}\right). \tag{5.9}$$

Comparison with Equation (5.2) shows that taking the variance of the squared length of the lateral component of mutation vectors into account reduces the estimate for the expected squared length of the lateral component by the second summand. This is reasonable as the lateral component is responsible for a negative contribution the fitness advantage associated with a mutation vector, and as preferably those mutation vectors with a short lateral component are selected for survival.

Using Equations (5.7) and (5.9) in Equation (5.1) yields an estimate for the quality gain of the $(\mu/\mu, \lambda)$-ES on the noisy sphere that is much improved as compared to Equation (5.4). However, the accuracy can be improved even further by accounting for the influence of the central component on the squared length of the progress vector. Still assuming that the contribution of the central component to the overall squared length can be ignored for the purpose of selection, the expected squared length of the central component of the progress vector is then simply given by Equation (D.34) and reads

$$\mathrm{E}\left[\langle z_A\rangle^2\right] = \frac{1}{\mu}\left(1 + \frac{h^{1,1}_{\mu,\lambda} + (\mu-1)h^{2,0}_{\mu,\lambda}}{1+\vartheta^2+\sigma^{*2}/2N}\right). \tag{5.10}$$

Finally, we take into account that the lateral component contributes $N-1$ rather than N terms to the squared length of the progress vector and therefore model the squared length of the progress vector as $\|\langle \mathbf{z}\rangle\|^2 = \langle z_A\rangle^2 + (N-1)\|\langle \mathbf{z}_B\rangle\|^2/N$. Using Equations (5.10) and (5.9) for $\mathrm{E}[\langle z_A\rangle^2]$ and $\mathrm{E}[\|\langle \mathbf{z}_B\rangle\|^2]$, our improved estimate of the expected squared length of the progress vector then reads

$$\mathrm{E}\left[\|\langle \mathbf{z}\rangle\|^2\right] = \frac{1}{\mu}\left(N + \frac{h^{1,1}_{\mu,\lambda} + (\mu-1)h^{2,0}_{\mu,\lambda}}{1+\vartheta^2+\sigma^{*2}/2N} - \frac{N-1}{N}\frac{\sigma^* h^{1,0}_{\mu,\lambda}}{\sqrt{1+\vartheta^2+\sigma^{*2}/2N}}\right). \tag{5.11}$$

From Equation (5.1) with Equations (5.7) and (5.11) it follows

$$\mathrm{E}[\Delta^*_{sphere}] = \frac{\sigma^* h^{1,0}_{\mu,\lambda}}{\sqrt{1+\vartheta^2+\sigma^{*2}/2N}} - \frac{\sigma^{*2}}{2\mu}\left(1 + \frac{1}{N}\frac{h^{1,1}_{\mu,\lambda} + (\mu-1)h^{2,0}_{\mu,\lambda}}{1+\vartheta^2+\sigma^{*2}/2N} - \frac{N-1}{N^2}\frac{\sigma^* h^{1,0}_{\mu,\lambda}}{\sqrt{1+\vartheta^2+\sigma^{*2}/2N}}\right) \tag{5.12}$$

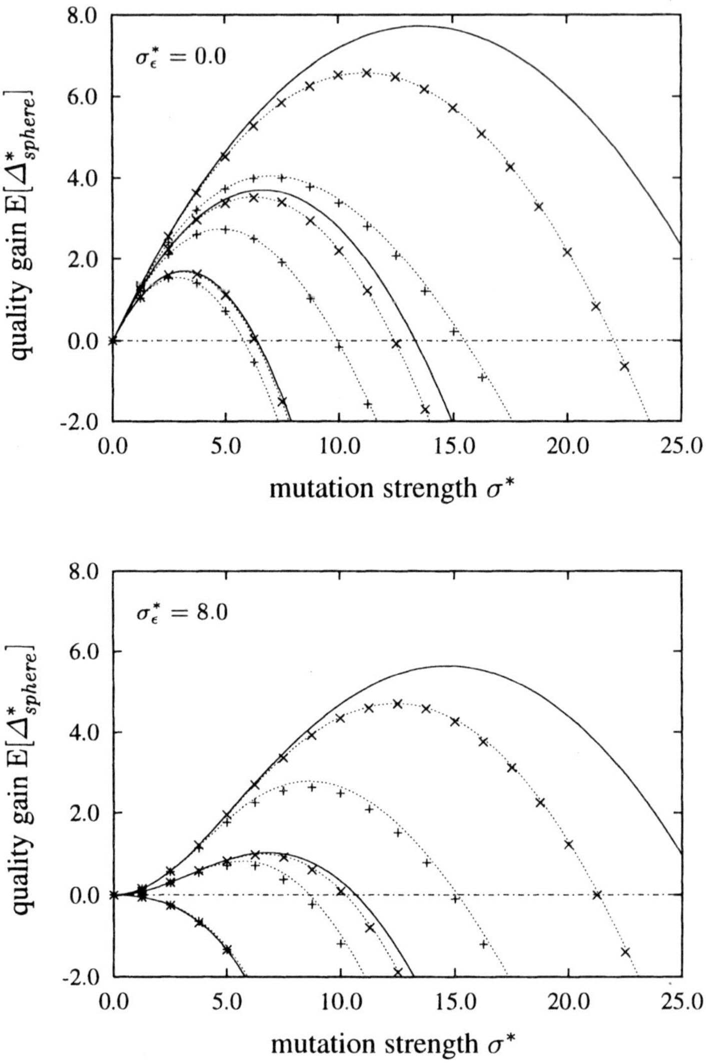

Figure 5.5. Normalized quality gain $E[\Delta^*_{sphere}]$ as a function of normalized mutation strength σ^* for several $(\mu/\mu, \lambda)$-ES. In the upper graph, the normalized noise strength is $\sigma_\epsilon^* = 0.0$, in the lower one it is $\sigma_\epsilon^* = 8.0$. The crosses correspond to, from bottom to top, measurements of a $(3/3, 10)$-ES, a $(6/6, 20)$-ES, and a $(12/12, 40)$-ES with search space dimensionalities $N = 40$ $(+)$ and $N = 400$ $(\times)$. The solid lines represent results from Equation (5.4). The dotted lines correspond to the improved approximation given by Equation (5.12).

as an improved estimate for the normalized quality gain of the $(\mu/\mu, \lambda)$-ES in the spherical environment. Note that for $N \to \infty$, Equation (5.4) is recovered. Figure 5.5 demonstrates the accuracy of the approximation thus obtained for several population sizes and noise strengths.

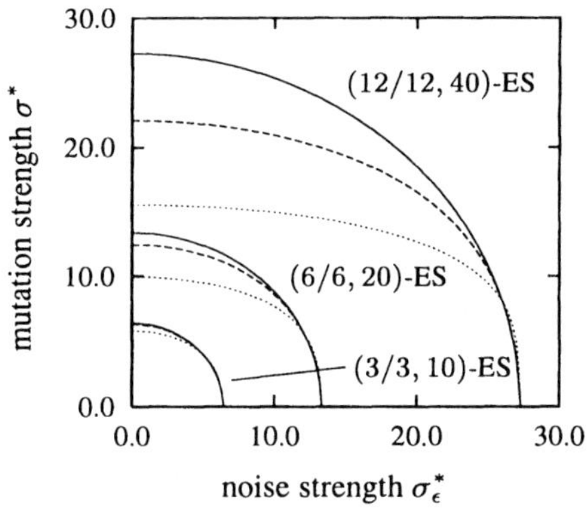

Figure 5.6. The normalized mutation strength σ^* up to which the quality gain is positive as a function of normalized noise strength σ_ϵ^*. The curves represent results obtained from Equation (5.12) for a $(3/3, 10)$-ES, a $(6/6, 20)$-ES, and a $(12/12, 40)$-ES for search space dimensionalities $N = 40$ (dotted lines), $N = 400$ (dashed lines), and $N = \infty$ (solid lines). The quality gain is positive below the respective curves and negative above.

2.2 Discussion

The estimate for the quality gain of the $(\mu/\mu, \lambda)$-ES thus obtained can now be used to study numerically optimal population sizes and efficiencies than can be achieved. Although not exact, the results can provide a good qualitative understanding of how the scaling properties of the $(\mu/\mu, \lambda)$-ES in finite-dimensional search spaces deviate from the ideal results obtained in Section 1. Closer empirical investigations confirm indeed that the inaccuracies of the improved approximation given by Equation (5.12) are small enough to be practically irrelevant for the results to be presented in the remainder of this section.

An important observation that is made when numerically evaluating the improved quality gain approximation just derived is that for finite N, mutation strengths cannot be chosen as large as Equation (5.4) indicates. Figure 5.6 has been obtained from Equation (5.12) and shows the maximal normalized mutation strength up to which positive quality gain can be achieved as a function of the normalized noise strength for different search space dimensionalities and population sizes. The quality gain is positive below the respective curves and negative above. It can be observed that while increasing the population size parameters μ and λ increases the region of positive quality gain, the accuracy of the approach from Section 1 deteriorates for fixed N. For the $(12/12, 40)$-ES with $N = 40$ and $\sigma_\epsilon^* = 0.0$, the range of mutation strengths that lead to positive quality gain is nearly halved as compared to the result based on Equation (5.4).

As a result of the reduced mutation strengths, genetic repair cannot be as effective as the results from Section 1 suggest. Figure 5.7 shows the dependence of the optimal efficiency η on the number of offspring per time step λ. It includes results from Section 1 that have already been illustrated in Figure 5.3 along with results from the improved approximation given by Equation (5.12). Two points are worth noting. First, it can be seen that for finite N, the efficiency of the $(\mu/\mu, \lambda)$-ES cannot be increased to the maximal value that can be achieved in the absence of noise by increasing the population size. Second, it

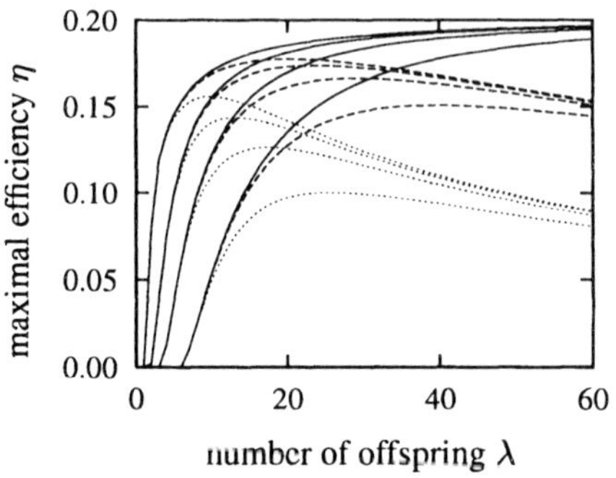

Figure 5.7. Maximal efficiency η as a function of the number of offspring per time step λ. The curves correspond to search space dimensionalities $N = 40$ (dotted lines), $N = 400$ (dashed lines), and $N = \infty$ (solid lines) with, from top to bottom, normalized noise strengths $\sigma_\epsilon^* = 0.0, 1.0, 2.0$, and 4.0.

can be seen that in contrast to the results obtained in Section 1, for finite search space dimensionality there is an optimal number of offspring above which the efficiency of the strategy declines. The choice of the parameter λ of the strategy becomes less critical with increasing noise strength as the maximum of the curves of Figure 5.7 becomes less pronounced.

Two important questions that follow immediately are how many offspring candidate solutions should be generated per time step to achieve optimal efficiency and how the efficiency that can be achieved compares with the ideal value of 0.202 found in Section 1. Both questions are answered in Figure 5.8 that has been obtained by numerically optimizing the efficiency of the $(\mu/\mu, \lambda)$-ES based on the improved approximation for the quality gain of the $(\mu/\mu, \lambda)$-ES on the noisy sphere that is given by Equation (5.12). The left hand graph shows that the optimal number of offspring increases with increasing noise strength and with increasing search space dimensionality and demonstrates that overall, optimal values of λ are rather moderate. The curves in the right hand graph show a clear decline of the maximal efficiency with increasing noise strength for finite N. However, except for very low normalized noise strengths, the maximal efficiency of the $(1+1)$-ES that has been derived in Chapter 3 and that is included in the graph for reference as well as the maximal efficiency of the $(1, \lambda)$-ES are far exceeded. Comparison with the right hand graph in Figure 4.4 shows that the efficiency of the (μ, λ)-ES is far exceeded as well. Thus, global intermediate recombination can qualitatively improve the local performance of evolution strategies in the presence of noise. While the improvement is not as great as the analysis from Section 1 suggests, it is nonetheless considerable.

3. Cumulative Mutation Strength Adaptation

In Chapter 2 we noted that mutation strength adaptation mechanisms are vital components of evolution strategies, and we briefly described cumulative mutation strength adaptation for isotropic mutations as one such mechanism. On the following pages we present an analysis of the performance of cumulative mutation strength adaptation on the noisy sphere. Making strong simplifications

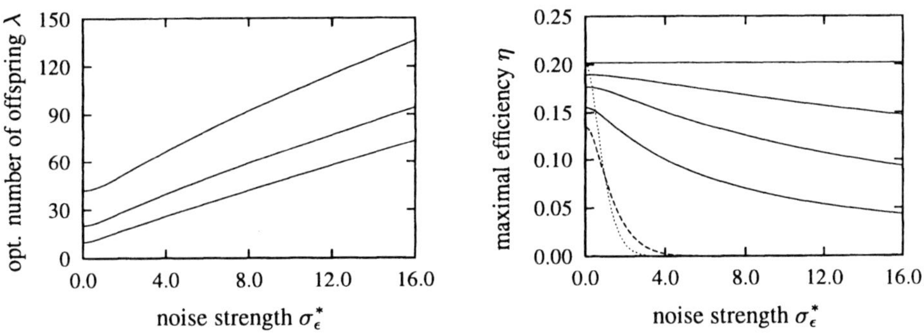

Figure 5.8. Optimal number of offspring per time step λ and maximal efficiency η as functions of normalized noise strength σ_ϵ^*. The solid curves correspond to, from bottom to top, search space dimensionalities $N = 40$, $N = 400$, and $N = 4000$. In the right hand graph, the limiting case $N = \infty$ is included as well. The dotted and dashed lines in the right hand graph represent the results for the $(1+1)$-ES derived in Chapter 3 and for the $(1, \lambda)$-ES that appears as a special case of Equation (5.5), respectively.

comparable to those made in the derivation of the simple model in Section 5 of Chapter 2, we arrive at simple laws that provide a good qualitative understanding of the capabilities of cumulative mutation strength adaptation in the presence of noise. We then use the improved estimate of the expected progress vector derived in Section 2 for arriving at quantitative results that allow for analyzing optimal population sizes and maximal efficiencies that can be achieved.

As for the $(1+1)$-ES without reevaluation of the parental fitness and for the (μ, λ)-ES, the local performance on the noisy sphere of the $(\mu/\mu, \lambda)$-ES with cumulative mutation strength adaptation cannot be evaluated by looking at single time steps only. Like the degree of overvaluation in case of the $(1+1)$-ES and like the central moments of the population in case of the (μ, λ)-ES, the mutation strength σ and the accumulated progress vector $\mathbf{s}$ of the $(\mu/\mu, \lambda)$-ES evolve in the course of the search and affect both each other and the quality gain that is achieved. The approach to analyzing cumulative mutation strength adaptation on the noisy sphere therefore closely parallels that pursued in Section 1.1 of Chapter 3 and in Sections 2 and 3 of Chapter 4. We consider quantities that after initialization effects have faded approach a time-invariant limit distribution. Important moments of that distribution can be obtained by computing expected values after a time step as functions of the respective values before the time step and demanding stationarity.

3.1 Computing the Expectations

As done for mutation vectors in Section 5 of Chapter 2 and for progress vectors in Section 1, the accumulated progress vector $\mathbf{s}$ in the spherical environment can be written as the sum of its central and lateral components, $\mathbf{s}_A$ and $\mathbf{s}_B$. We write s_A for the signed length of the central component of the accumulated progress vector. That is, s_A equals the length of the central component if the accumulated progress vector points towards the origin and it equals the negative thereof if it points away from it. For symmetry reasons, the direction of the lateral component of the accumulated progress vector is without relevance for the performance of the strategy. As the progress vector, the accumulated progress vector is well characterized by the signed length s_A of its central component and by its overall squared length $||\mathbf{s}||^2$. These two quantities together with the normalized mutation strength σ^* form the state variables the time-invariant limit distribution of we seek to learn about.

For notational simplicity, we write $\mathbf{x}$ and $\mathbf{z}$ instead of $\langle\mathbf{x}\rangle$ and $\langle\mathbf{z}\rangle$ for the centroid of the population and for the progress vector, respectively. Again, due to the spherical symmetry, it can without loss of generality be assumed that $\hat{\mathbf{x}} - \mathbf{x}^{(t)} = (R^{(t)}, 0, \ldots, 0)^{\mathrm{T}}$ and therefore that the central and lateral components of the progress vector and of the accumulated progress vector at time t are $\mathbf{z}_A^{(t)} = (z_A^{(t)}, 0, \ldots, 0)^{\mathrm{T}}$, $\mathbf{z}_B^{(t)} = (0, z_2^{(t)}, \ldots, z_N^{(t)})^{\mathrm{T}}$, $\mathbf{s}_A^{(t)} = (s_A^{(t)}, 0, \ldots, 0)^{\mathrm{T}}$, and $\mathbf{s}_B^{(t)} = (0, s_2^{(t)}, \ldots, s_N^{(t)})^{\mathrm{T}}$, respectively.

Squared Length of the Accumulated Progress Vector

Using Equation (2.3), the squared length of the accumulated progress vector at time $t+1$ is

$$\begin{aligned} ||\mathbf{s}^{(t+1)}||^2 &= \sum_{i=1}^{N} \left((1-c)s_i^{(t)} + \sqrt{c(2-c)}\sqrt{\mu} z_i^{(t)} \right)^2 \\ &= (1-c)^2 ||\mathbf{s}^{(t)}||^2 + 2(1-c)\sqrt{c(2-c)}\sqrt{\mu} \sum_{i=1}^{N} s_i^{(t)} z_i^{(t)} \\ &\qquad + c(2-c)\mu ||\mathbf{z}^{(t)}||^2. \end{aligned}$$

Taking the expected value it follows that

$$\begin{aligned} \mathrm{E}\left[||\mathbf{s}^{(t+1)}||^2 \right] &= (1-c)^2 ||\mathbf{s}^{(t)}||^2 + 2(1-c)\sqrt{c(2-c)}\sqrt{\mu} s_A^{(t)} \mathrm{E}\left[z_A^{(t)} \right] \\ &\qquad + c(2-c)\mu \mathrm{E}\left[||\mathbf{z}^{(t)}||^2 \right], \end{aligned} \tag{5.13}$$

where we have used the fact that $\mathrm{E}[z_i^{(t)}] = 0$ for $i \neq 1$ due to symmetry considerations.

Signed Length of the Central Component of the Accumulated Progress Vector

The signed length of the central component of the accumulated progress vector equals the inner product of the accumulated progress vector with a vector of length unity pointing from the centroid of the population to the optimizer. Thus, using Equations (2.2) and (2.3), the signed length of the central component of the accumulated progress vector at time $t+1$ is

$$\begin{aligned} s_A^{(t+1)} &= \left((1-c)\mathbf{s}^{(t)} + \sqrt{c(2-c)}\sqrt{\mu}\mathbf{z}^{(t)}\right)^{\mathrm{T}} \frac{\left(\hat{\mathbf{x}} - \mathbf{x}^{(t)}\right) - \sigma^{(t)}\mathbf{z}^{(t)}}{R^{(t+1)}} \\ &= \frac{R^{(t)}}{R^{(t+1)}} \left[(1-c)\left(s_A^{(t)} - \frac{\sigma^{*(t)}}{N}\sum_{i=1}^{N} s_i^{(t)} z_i^{(t)}\right)\right. \\ &\qquad \left. + \sqrt{c(2-c)}\sqrt{\mu}\left(z_A^{(t)} - \frac{\sigma^{*(t)}}{N}\|\mathbf{z}^{(t)}\|^2\right)\right], \end{aligned}$$

where the fact that the direction from the centroid of the population to the optimizer at time t is aligned with the x_1-axis has been used. As for the quotient $R^{(t)}/R^{(t+1)}$, Taylor expansion shows that

$$\begin{aligned} \frac{R^{(t)}}{R^{(t+1)}} &= \left(1 - \frac{2\Delta^{*(t)}_{sphere}}{N}\right)^{-1/2} \\ &= 1 + \frac{\Delta^{*(t)}_{sphere}}{N} + \frac{3}{2}\left(\frac{\Delta^{*(t)}_{sphere}}{N}\right)^2 + \ldots. \end{aligned}$$

Let us make the assumption that in the present context, for sufficiently large N, we can replace $R^{(t)}/R^{(t+1)}$ by unity. While there is currently no justification for this assumption, it will appear reasonable in the light of the results to be obtained below. Then, taking the expected value, all terms but the first in the sum ranging from 1 to N vanish as $\mathrm{E}[z_i^{(t)}] = 0$ for $i \neq 1$. Moreover, the remaining term $\sigma^{*(t)}\mathrm{E}[z_A^{(t)}]/N$ can be neglected as compared to unity and it follows

$$\begin{aligned} \mathrm{E}\left[s_A^{(t+1)}\right] &\approx (1-c)s_A^{(t)} \\ &\quad + \sqrt{c(2-c)}\sqrt{\mu}\left(\mathrm{E}\left[z_A^{(t)}\right] - \frac{\sigma^{*(t)}}{N}\mathrm{E}\left[\|\mathbf{z}^{(t)}\|^2\right]\right), \end{aligned} \tag{5.14}$$

for the expected signed length of the accumulated progress vector at time $t+1$.

Normalized Mutation Strength

From Equations (2.4) and (2.12), it follows that the normalized mutation strength at time $t+1$ is

$$\sigma^{*(t+1)} = \sigma^{*(t)}\frac{R^{(t)}}{R^{(t+1)}}\exp\left(\frac{\|\mathbf{s}^{(t+1)}\|^2 - N}{2DN}\right).$$

Expanding both the quotient $R^{(t)}/R^{(t+1)}$ and the exponential function into Taylor series yields

$$\sigma^{*(t+1)} = \sigma^{*(t)}\left(1 + \frac{\Delta^{*(t)}_{sphere}}{N} + \ldots\right)\left(1 + \frac{\|\mathbf{s}^{(t+1)}\|^2 - N}{2DN} + \ldots\right).$$

Multiplying out, taking the expectation, and neglecting all terms but the first few yields

$$\mathrm{E}\left[\sigma^{*(t+1)}\right] \approx \sigma^{*(t)}\left(1 + \frac{\mathrm{E}[\Delta^{*(t)}_{sphere}]}{N} + \frac{\mathrm{E}\left[\|\mathbf{s}^{(t+1)}\|^2\right] - N}{2DN}\right) \tag{5.15}$$

for the expected normalized mutation strength at time $t+1$. Again, a justification for neglecting the terms that have been omitted can only be obtained a posteriori.

3.2 Evaluating the Performance

In the following paragraphs, we obtain an estimate of the expected accumulated progress vector by demanding stationarity. We then set out and define and calculate the adaptation response and the target mutation strength of the strategy. Finally, we analyze the dynamic performance of cumulative mutation strength adaptation and investigate optimal population sizes and efficiencies that can be attained.

Determining the Accumulated Progress Vector

The right hand sides of Equations (5.13), (5.14), and (5.15) depend on $\|\mathbf{s}\|^2$, s_A, and σ^* in such a way that determination of the moments by means of integration over the limit distribution of the quantities as done in Section 2 of Chapter 4 is precluded. Instead, as in Section 3 of that chapter, we ignore fluctuations and look for those values of the state variables that form a fixed point of the mapping given by the equations. That is, we solve the system of equations

$$\left.\begin{aligned}
\|\mathbf{s}\|^2 &= (1-c)^2\|\mathbf{s}\|^2 + 2(1-c)\sqrt{c(2-c)}\sqrt{\mu}s_A\mathrm{E}\left[z_A\right] \\
&\qquad + c(2-c)\mu\mathrm{E}\left[\|\mathbf{z}\|^2\right] \\
s_A &= (1-c)s_A + \sqrt{c(2-c)}\sqrt{\mu}\left(\mathrm{E}\left[z_A\right] - \frac{\sigma^*}{N}\mathrm{E}\left[\|\mathbf{z}\|^2\right]\right) \\
\sigma^* &= \sigma^*\left(1 + \frac{\mathrm{E}[\Delta^*_{sphere}]}{N} + \frac{\|\mathbf{s}\|^2 - N}{2DN}\right)
\end{aligned}\right\} \tag{5.16}$$

for $\|\mathbf{s}\|^2$, s_A, and σ^* and hope that the resulting values are good estimates for the expected values of the respective quantities.

With the expected values $\mathrm{E}[\|\mathbf{z}\|^2]$, $\mathrm{E}[z_A]$, and $\mathrm{E}[\Delta^*_{sphere}]$ given by Equations (5.2), (5.3), and (5.4), respectively, it is easily verified that solving the second equation for the signed length of the central component of the accumulated progress vector yields

$$s_A = \frac{\sqrt{c(2-c)}}{c}\sqrt{\mu} h^{1,0}_{\mu,\lambda}\left(\frac{1}{\sqrt{1+\vartheta^2}} - \frac{\sigma^*}{\mu h^{1,0}_{\mu,\lambda}}\right). \tag{5.17}$$

Using this result in the first of the equations of the system to be solved it follows

$$\|\mathbf{s}\|^2 = N + \frac{2(1-c)}{c}\frac{\mu {h^{1,0}_{\mu,\lambda}}^2}{\sqrt{1+\vartheta^2}}\left(\frac{1}{\sqrt{1+\vartheta^2}} - \frac{\sigma^*}{\mu h^{1,0}_{\mu,\lambda}}\right) \tag{5.18}$$

for the squared length of the accumulated progress vector. While being inexact due to the simplifications made in their derivation, for large N, Equations (5.17) and (5.18) do provide a good qualitative understanding of the behavior of cumulative mutation strength adaptation on the noisy sphere.

Static Analysis

Before considering the normalized mutation strength that cumulative mutation strength adaptation realizes, let us first analyze the static behavior of the strategy. For that purpose, let us define the *logarithmic adaptation response*

$$\Delta^{(t)}_\sigma = \log\left(\frac{\sigma^{(t+1)}}{\sigma^{(t)}}\right) \tag{5.19}$$

and its normalization

$$\Delta^*_\sigma = \Delta_\sigma \frac{cDN}{1-c}$$

to describe the performance of the mutation strength adaptation scheme. Note that Beyer [22] offered a different definition for a self-adaptation response that is similar in spirit. The logarithmic adaptation response quantifies how the mutation strength adaptation mechanism responds to an ill-adapted mutation strength. Ideally, it is positive for mutation strengths that are too small and negative for mutation strengths that are too large. The root of the logarithmic adaptation response determines the *target mutation strength* of the strategy.

From Equations (2.4) and (5.18), we obtain the estimate

$$\Delta^*_\sigma = \frac{\mu {h^{1,0}_{\mu,\lambda}}^2}{\sqrt{1+\vartheta^2}}\left(\frac{1}{\sqrt{1+\vartheta^2}} - \frac{\sigma^*}{\mu h^{1,0}_{\mu,\lambda}}\right), \tag{5.20}$$

for the normalized logarithmic adaptation response of cumulative mutation strength adaptation on the noisy sphere. Root finding shows that the target

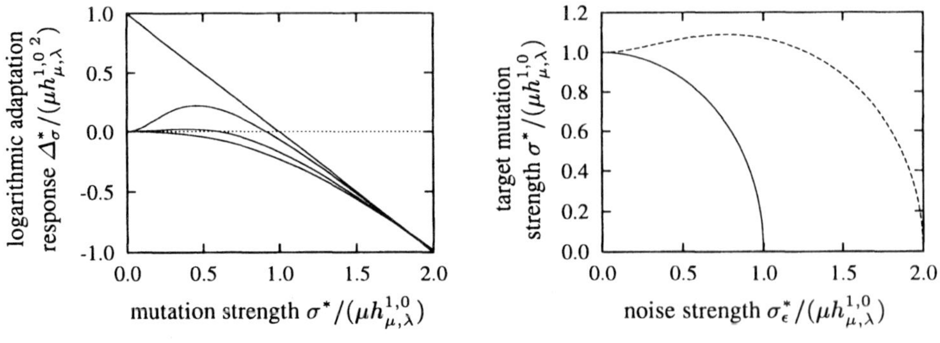

Figure 5.9. Normalized logarithmic adaptation response Δ_σ^* as a function of normalized mutation strength σ^* and normalized target mutation strength σ^* as a function of the normalized noise strength σ_ϵ^*. Note the scaling of the axes. The lines in the left hand graph have been obtained from Equation (5.20) for, from top to bottom, normalized noise strengths $\sigma_\epsilon^* / (\mu h_{\mu,\lambda}^{1,0}) = 0.0$, 0.4, 0.8, and 1.2. In the right hand graph, the solid line corresponds to the target mutation strength given by Equation (5.21), the dashed line to the optimal mutation strength that has been depicted in Figure 5.2.

mutation strength is

$$\sigma^* = \mu h_{\mu,\lambda}^{1,0} \sqrt{1 - \left(\frac{\sigma_\epsilon^*}{\mu h_{\mu,\lambda}^{1,0}} \right)^2}. \tag{5.21}$$

Figure 5.9 shows the normalized logarithmic adaptation response given by Equation (5.20) as a function of the normalized mutation strength and the normalized target mutation strength given by Equation (5.21) as a function of the normalized noise strength. At least two things can be learned from the figure. First, the right hand graph shows that for nonzero noise strength, the target mutation strength is below the mutation strength that maximizes the quality gain of the strategy. For normalized noise strengths exceeding $\mu h_{\mu,\lambda}^{1,0}$, the target mutation strength is zero even though positive quality gain could be achieved with nonzero mutation strengths. Second, the left hand graph shows that in the presence of noise, for mutation strengths significantly below their optimal values, the logarithmic adaptation response and therefore the tendency towards higher mutation strengths is very small. This is intuitively clear as in Section 2 of Chapter 2 we noted that cumulative mutation strength adaptation attempts to achieve that consecutive progress vectors are uncorrelated. Small steps carry little information. In the presence of noise, that information is almost entirely hidden and correlations between consecutive progress vectors disappear as steps are increasingly random. The strategy thus sees no need to increase the mutation strength much even though significantly higher mutation strengths would be required to achieve a better signal-to-noise ratio. This insight sheds new light on the postulate suggested by Beyer and Deb [30] that in flat fitness landscapes, mutation strength adaptation schemes should systematically increase the mutation strength. In the presence of noise, this may be especially useful as fitness landscapes may appear to be flat due to a low signal-to-noise ratio, and as operating at higher mutation strengths might make more reliable information available to the strategy.

Dynamic Analysis

The target mutation strength is not the mutation strength that is actually realized by the mutation strength adaptation mechanism. As the distance R to the optimizer continually changes, and as adaptation to the target mutation strength is not instantaneous, the mutation strength that is actually realized is always "behind". An estimate of that mutation strength can be computed by solving the system of Equations (5.16) for σ^*. It is easily verified that using Equations (5.4), (5.17), and (5.18), the third one of the equations can be transformed into

$$\frac{1-c}{cD} + \left(1 - \frac{1-c}{cD}\right)\sqrt{\left(\frac{\sigma^*}{\mu h_{\mu,\lambda}^{1,0}}\right)^2 + \left(\frac{\sigma_\epsilon^*}{\mu h_{\mu,\lambda}^{1,0}}\right)^2}$$
$$= \frac{1}{2}\left(\frac{\sigma^*}{\mu h_{\mu,\lambda}^{1,0}}\right)^2 + \frac{1}{2}\left(\frac{\sigma_\epsilon^*}{\mu h_{\mu,\lambda}^{1,0}}\right)^2.$$

For the settings of c and D suggested by Hansen [42], i.e. $c = 1/\sqrt{N}$ and $D = \sqrt{N}$, the first term on the left hand side tends to unity as N increases, and the second term asymptotically vanishes. Therefore, solving for the normalized mutation strength that the mutation strength adaptation mechanism realizes yields

$$\sigma^* = \mu h_{\mu,\lambda}^{1,0}\sqrt{2 - \left(\frac{\sigma_\epsilon^*}{\mu h_{\mu,\lambda}^{1,0}}\right)^2}. \tag{5.22}$$

According to Equation (5.4), the quality gain achieved with this mutation strength is

$$\mathrm{E}[\Delta_{sphere}^*] = \frac{\sqrt{2}-1}{2}\mu h_{\mu,\lambda}^{1,0\,2}\left(2 - \left(\frac{\sigma_\epsilon^*}{\mu h_{\mu,\lambda}^{1,0}}\right)^2\right). \tag{5.23}$$

Both the mutation strength given by Equation (5.22) and the quality gain given by Equation (5.23) are shown as functions of the normalized noise strength in Figure 5.10. It can be seen that even though the target mutation strength of cumulative mutation strength adaptation is optimal for zero noise strength and too small in the presence of noise, the dynamics of the process lead to

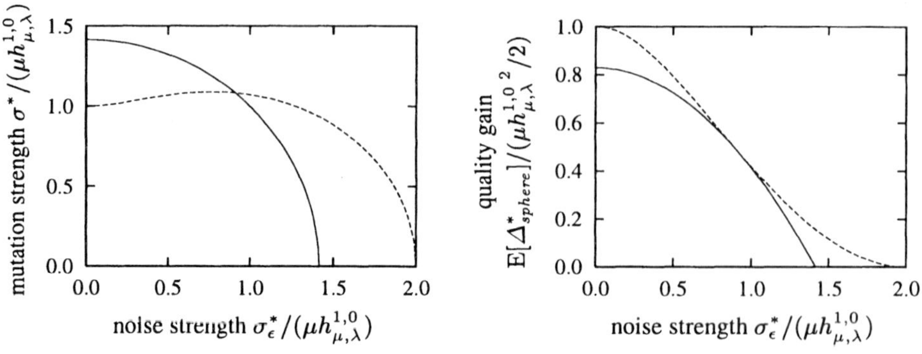

Figure 5.10. Normalized mutation strength σ^* and normalized quality gain $\mathrm{E}[\Delta^*_{sphere}]$ as functions of the normalized noise strength σ^*_ϵ. Note the scaling of the axes. The solid lines correspond to the $(\mu/\mu, \lambda)$-ES with cumulative mutation strength adaptation and have been obtained from Equations (5.22) and (5.23). The dashed lines represent the optimal values that have been depicted in Figure 5.2.

mutation strengths that are too large for normalized noise strengths of up to about $0.91\mu h^{1,0}_{\mu,\lambda}$ and too small for noise strengths above this value. Positive quality gain is achieved up to a normalized noise strength of $\sqrt{2}\mu h^{1,0}_{\mu,\lambda}$.

Population Sizing

As noted in Section 2, the accuracy of the simple model for the spherical environment that was introduced in Section 5 of Chapter 2 and that has been used in the present section so far is rather bad. While sufficient for obtaining a qualitative understanding of the performance of cumulative mutation strength adaptation in the presence of noise, the approximation considered thus far is too crude for studying questions like the problem of determining optimal population sizes. In what follows, we therefore use the improved approximation for the progress vector and the quality gain from Section 2. Moreover, we do not ignore the terms that had been neglected in the derivations leading to Equations (5.14) and (5.15). However, we do still assume that progress vector and quality gain can be replaced by their expected values, and we still look for a fixed point of the iterated map rather than considering the distribution functions of the quantities. Replacing Equations (5.16), the resulting system of equations

$$\|\mathbf{s}\|^2 = (1-c)^2\|\mathbf{s}\|^2 + 2(1-c)\sqrt{c(2-c)\mu}s_A\mathrm{E}\left[z_A\right] + c(2-c)\mu\mathrm{E}\left[\|\mathbf{z}\|^2\right]$$

$$s_A\sqrt{1+\frac{2}{N}\mathrm{E}[\Delta^*_{sphere}]} = (1-c)s_A\left(1-\frac{\sigma^*}{N}\mathrm{E}[z_A]\right) + \sqrt{\mu c(2-c)}\left(\mathrm{E}\left[z_A\right]-\frac{\sigma^*}{N}\mathrm{E}\left[\|\mathbf{z}\|^2\right]\right)$$

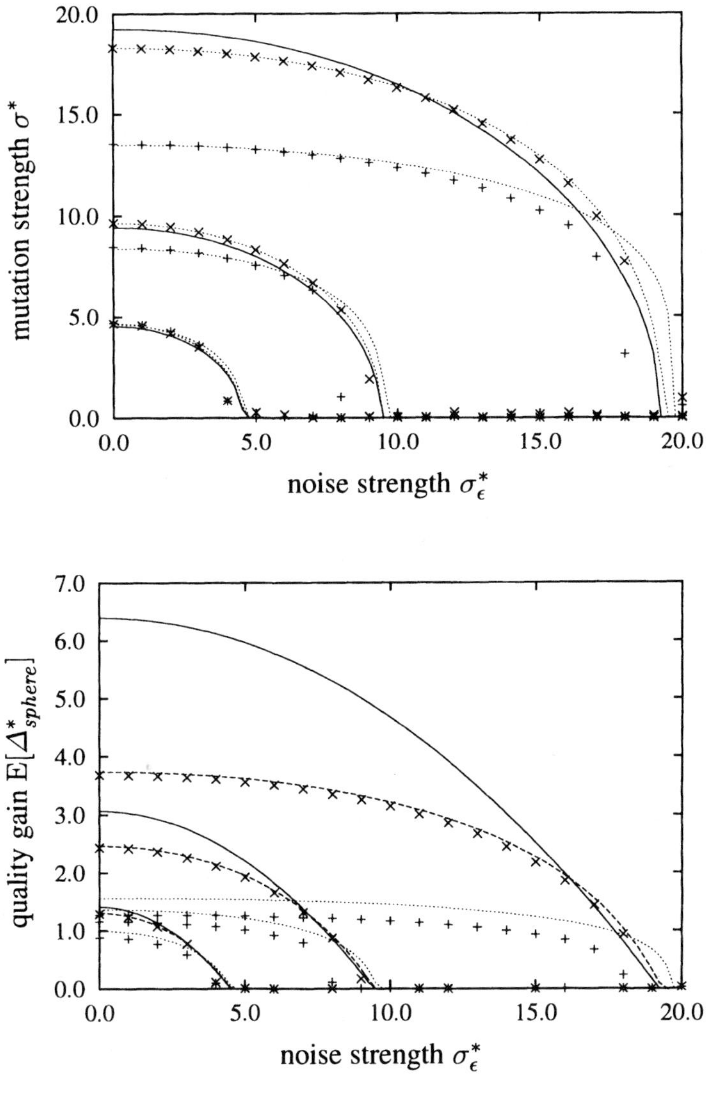

Figure 5.11. Normalized mutation strength σ^* and normalized quality gain $E[\Delta^*_{sphere}]$ as functions of the normalized noise strength σ_ϵ^*. The crosses correspond to, from bottom to top, measurements of a $(3/3, 10)$-ES, a $(6/6, 20)$-ES, and a $(12/12, 40)$-ES with cumulative mutation strength adaptation at search space dimensionalities $N = 40$ $(+)$ and $N = 400$ $(\times)$. The solid lines represent the estimates for infinite search space dimensionality given by Equations (5.22) and (5.23) that are also depicted in Figure 5.10. The dotted lines correspond to the improved approximation that takes some N-dependent terms into account.

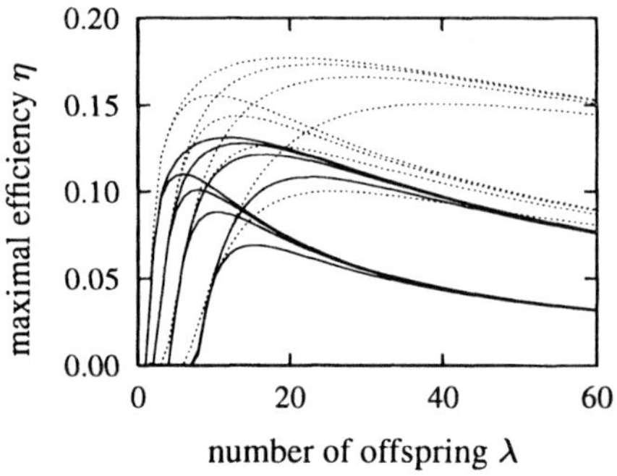

Figure 5.12. Maximal efficiency η as a function of the number of offspring per time step λ for, from top to bottom, normalized noise strengths $\sigma_\epsilon^* = 0.0, 1.0, 2.0$, and 4.0. The curves correspond to search space dimensionalities $N = 40$ (lower four curves) and $N = 400$ (upper four curves). The solid lines represent results for cumulative mutation strength adaptation, the dotted lines are the values for optimal mutation strength.

$$\sqrt{1 + \frac{2}{N}\mathrm{E}[\Delta^*_{sphere}]} = \exp\left(\frac{||\mathbf{s}||^2 - N}{2DN}\right)$$

where $\mathrm{E}[||\mathbf{z}||^2]$, $\mathrm{E}[z_A]$, and $\mathrm{E}[\Delta^*_{sphere}]$ are given by Equations (5.11), (5.7), and (5.12), can be solved numerically for the normalized mutation strength, and that normalized mutation strength can be used to obtain an estimate of the normalized quality gain using Equation (5.12).

Figure 5.11 compares the estimates obtained with empirical measurements of the normalized mutation strength and the normalized quality gain of the $(\mu/\mu, \lambda)$-ES averaged over many time steps on noisy spheres with search space dimensionalities $N = 40$ and $N = 400$. It can be seen that the accuracy of the estimates for the mutation strength is very good except for high normalized noise strength. This inaccuracy is tolerable as the $(\mu/\mu, \lambda)$-ES optimally uses population sizes large enough to guarantee that the region in which the inaccuracies occur is avoided. The lower graph shows that the estimate for the quality gain is not quite as good. This has two reasons: fluctuations of the mutation strength and the limited accuracy of the quality gain law for high mutation strengths. Recall that for moderate noise strengths cumulative mutation strength adaptation realizes mutation strengths exceeding the optimal values. However, it can also be seen that the inaccuracy of the estimate for the quality gain decreases with increasing N.

On the basis of the estimates thus obtained, optimal population sizes and efficiencies can be determined. While not exact, the results allow for a comparison with results that have been obtained in Section 2 under the assumption of optimally adapted mutation strength. The qualitative differences in the results are reflected correctly. Figure 5.12 shows the maximal efficiency η as a function of the number of offspring per time step λ. Little surprisingly, it can be seen that the values obtained for optimally adapted mutation strength cannot be attained. Figure 5.13 shows the optimal number of offspring per time step λ and the maximal efficiency as functions of the normalized noise strength σ_ϵ^*. It can be seen that using cumulative mutation strength adaptation, optimal pop-

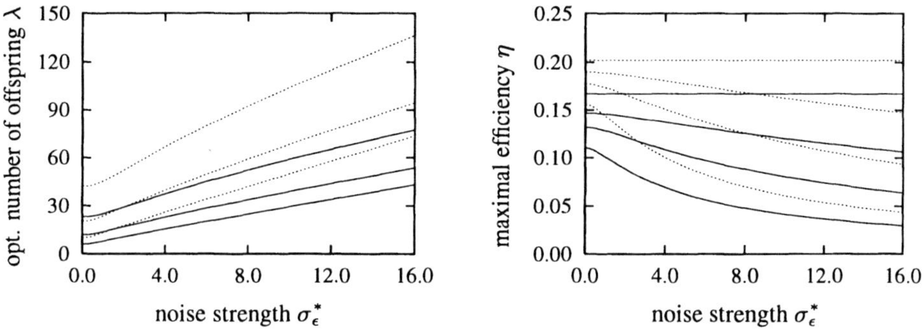

Figure 5.13. Optimal number of offspring per time step λ and maximal efficiency η as functions of the normalized noise strength σ_ϵ^*. The curves correspond to, from bottom to top, search space dimensionalities $N = 40$, $N = 400$, and $N = 4000$. In the right hand graph, the limiting case $N = \infty$ is included as well. The solid curves depict the results for cumulative mutation strength adaptation. The dotted lines assume optimally adapted mutation strength and have also been depicted in Figure 5.8.

ulation sizes are below the already moderate values computed in Section 2. It can also be seen that the efficiencies that can be attained using cumulative mutation strength adaptation are not very much smaller than the optimal ones, and thus that cumulative mutation strength adaptation works quite well on the noisy sphere.

Chapter 6

COMPARING APPROACHES TO NOISY OPTIMIZATION

In the previous chapters, we have studied the performance of various evolution strategies on the noisy sphere. We have seen the beneficial effects of overvaluation that follow from failure to reevaluate parental fitness along with the problems for success probability-based mutation strength adaptation rules that result. We have investigated the gain that can be achieved by using populations of candidate solutions that are distributed in search space and the benefits of genetic repair in the presence of noise. With the analysis of the behavior of the $(\mu/\mu, \lambda)$-ES with cumulative mutation strength adaptation in Section 3 of Chapter 5, the dynamics of a complete evolution strategy in the presence of noise are now understood. However, evolution strategies are by no means the only algorithms used for optimization in the presence of noise, and nothing has yet been said with regard to their capabilities relative to those competing approaches. It is therefore the goal of this chapter to contrast the performance in the presence of noise of evolution strategies with that of other algorithms that are used frequently or devised explicitly for noisy optimization.

Comparisons of the performance of evolutionary algorithms and of other optimization strategies in the presence of noise have been made before. In an early work, Gurin and Rastrigin [40] published a theoretical comparison of what they refer to as a random search method with a gradient strategy on a linear fitness function. Their random search method could be characterized as a $(1 + 1)$-ES with reevaluation of the parental objective function value and with rescaled mutations that are uniformly distributed on a hypersphere. The analysis assumes that the step lengths for the two strategies are identical and finds that due to the high cost involved in approximating the gradient by means of forward differencing the random search strategy is superior for all but the smallest values of N. More recently, both Nissen and Propach [59, 60] and Rana, Whitley, and Cogswell [64] numerically compared the performance of several point-based and population-based optimization strategies on a number of noisy objective functions. They found the population-based methods to be generally more robust with regard to the effects of noise. Finally, both Kelley [50] and Torczon [86] suggested new algorithms for noisy optimization and empirically compared their performance with that of several standard strategies for a number of objective functions.

In the comparison to be presented here, we restrict ourselves to evaluating the performance of different search strategies on the noisy sphere with fitness-proportionate noise strength rather than considering a larger range of fitness environments. The sphere is arguably the most simple nontrivial function that can be considered, and it seems plausible that an optimization strategy that fails on the sphere is unlikely to succeed when facing more difficult tasks. The sphere is also an interesting function to consider because of the transparency of the results it affords. As seen before, the assumption of fitness-proportionate noise strength results in scale invariance of the environment. We average the performance of the strategies we consider over long periods of time and after enough time has passed for initialization effects to have faded. Both initialization procedures and termination criteria are without relevance for our computer experiments. Thus, the influence of the normalized noise strength and of the search space dimensionality are isolated and scaling properties can be derived visually from our performance plots. The results can be interpreted easily and reveal interesting information with regard to the strengths and shortcomings of the strategies.

In Section 1, we outline the optimization strategies the performance of which we compare with that of the evolution strategies. Section 2 consists of a detailed description of the conditions under which the computer experiments are carried out and then summarizes and interprets their results.

1. The Competitors

For our comparison, we assume that only (noisy) objective function values are available to the search strategies. In particular, no derivatives are known, and search algorithms that make use of gradient information need to approximate derivatives for example by means of finite differences. Providing a classification of search methods that accommodates all of the strategies that have been suggested for this problem is a difficult task as can be seen from the discussion of what constitutes a direct search strategy by Trosset [88]. Useful classification schemes can be found in papers by Powell [62] and by Lewis, Torczon, and Trosset [53]. An extensive overview of optimization algorithms has also been compiled by Schwefel [77].

The number of search strategies that can be considered in any comparison such as that to be presented below is necessarily limited. In our selection of algorithms, we choose strategies that are of historical importance and in widespread use as well as some promising recent developments. We only consider algorithms that are adaptive in that parameters such as step lengths are changed in response to results of objective function evaluations rather than being controlled by fixed schedules. As due to the nature of our experiments there is no need to consider initialization procedures or termination criteria, we do not discuss those in what follows.

Many of the search methods in use today can be traced back to either the approach of *response surface methodology* or to the field of *stochastic approximation*. The foundations of response surface methodology were established by Box and Wilson [34] who were concerned with minimizing an unknown quadratic objective function disturbed by random noise of constant noise strength. They proposed constructing a local linear or quadratic model of the objective function by performing experiments in the neighborhood of the current iterate, and to take a step in the direction of steepest descent as derived from this model. According to Torczon and Trosset [87], the designs employed in these experiments would become the patterns in pattern search methods. Response surface methodology thus is a direct precursor of methods such as those of Hooke and Jeeves [47] (Section 1.1), Spendley, Hext and Himsworth [83], Nelder and Mead [58] (Section 1.2), and Torczon [86] (Section 1.3). Stochastic approximation on the other hand dates back to work of Robbins and Monro [69] and Kiefer and Wolfowitz [52]. The latter authors suggested the use of finite differencing for obtaining an approximation to the gradient of an unknown, noisy function, and to proceed in direction of this approximate gradient. The algorithms of Gilmore and Kelley [38] (Section 1.4) and Spall [78] (Section 1.5) are derived from this approach. Finally, the evolutionary gradient search method of Salomon [74, 75] (Section 1.6) is inspired both by gradient and by evolutionary methods.

1.1 Direct Pattern Search

The direct pattern search algorithm of Hooke and Jeeves [47] is an early example of a direct search strategy. The state of the strategy at time t is described by a base point $\mathbf{x}^{(t)} \in I\!R^N$, a vector $\mathbf{d}^{(t)} \in I\!R^N$ that equals the most recently taken step, and step length $h^{(t)}$. An iteration of the direct pattern search algorithm consists of a pattern step and a sequence of exploratory steps. The pattern step results in intermediate point $\mathbf{y}^{(t)} = \mathbf{x}^{(t)} + \mathbf{d}^{(t)}$, thus duplicating the most recently taken step in the hope that it may speed up the search. Note that if the most recent step was unsuccessful, $\mathbf{d}^{(t)}$ equals the zero vector and effectively no pattern step is being made. The sequence of exploratory moves starts at $\mathbf{y}^{(t)}$ by successively taking steps of length h along the axes $\mathbf{e}_i$, $i = 1, \ldots, N$, of the coordinate system. If such a step leads to an improved objective function value, it is accepted. Otherwise, taking a step in the opposite direction $-\mathbf{e}_i$ is attempted. In our implementation of the algorithm of Hooke and Jeeves we have made use of the improvement suggested by Bell and Pike [17] that aims at reducing the number of objective function evaluations in the sequence of exploratory steps by remembering the most recently taken of the two possible directions as the more promising one.

Only after exploratory moves along all axes of the coordinate system have been attempted and a point $\mathbf{z}^{(t)}$ has been reached is the objective function value $f(\mathbf{z}^{(t)})$ compared with $f(\mathbf{x}^{(t)})$ to decide whether the overall step is to be taken. If it is, then $\mathbf{d}^{(t+1)}$ is set to $\mathbf{z}^{(t)} - \mathbf{x}^{(t)}$ and the new base point $\mathbf{x}^{(t+1)}$ is set to be $\mathbf{z}^{(t)}$. If the overall step is rejected, then $\mathbf{d}^{(t+1)}$ is set to be zero. In addition, if $\mathbf{d}^{(t)}$ already equaled zero before the step, then the step length h is halved. Thus, a step of the direct pattern search algorithm involves between $N + 1$ and $2N + 1$ objective function evaluations.

1.2 Simplex Method

Based on an earlier numerical optimization scheme by Spendley, Hext, and Himsworth [83], Nelder and Mead [58] in 1965 devised a simplex method for function minimization. A simplex is the convex hull of $N + 1$ points in $I\!R^N$, where the points satisfy the nondegeneracy condition that the volume of the hull is nonzero. In every iteration of the algorithm, the vertices $\mathbf{x}_1, \ldots, \mathbf{x}_{N+1}$ of the current simplex are ordered according to their (noisy) objective function values such that $f(\mathbf{x}_1) \leq f(\mathbf{x}_2) \leq \cdots \leq f(\mathbf{x}_{N+1})$. We refer to $\mathbf{x}_1$ as the best point and to $\mathbf{x}_{N+1}$ as the worst point. Moreover, we write $\langle \mathbf{x} \rangle = \sum_{i=1}^{N} \mathbf{x}_i / N$ for the centroid of all points but the worst. The simplex method attempts to replace the current worst vertex by a new vertex that is generated by a reflection, by an expansion, or by a contraction. Only in case this fails is a shrink step carried out. The exact sequence of operations is as follows:

Reflect: A reflection point

$$\mathbf{x}_{\mathit{reflect}} = \langle \mathbf{x} \rangle + \alpha(\langle \mathbf{x} \rangle - \mathbf{x}_{N+1})$$

is computed and its objective function value is evaluated. If $f(\mathbf{x}_1) \leq f(\mathbf{x}_{\mathit{reflect}}) < f(\mathbf{x}_N)$, i.e. if the reflected point improves on the worst point but is not better than the best point so far, then $\mathbf{x}_{N+1}$ is replaced by $\mathbf{x}_{\mathit{reflect}}$ and the iteration is terminated.

Expand: If $f(\mathbf{x}_{\mathit{reflect}}) < f(\mathbf{x}_1)$, i.e. if the reflected point is better than the best point so far, then an expansion point

$$\mathbf{x}_{\mathit{expand}} = \langle \mathbf{x} \rangle + \gamma(\mathbf{x}_{\mathit{reflect}} - \langle \mathbf{x} \rangle)$$

is computed and its objective function value is evaluated. The worst point $\mathbf{x}_{N+1}$ is replaced by the better of $\mathbf{x}_{\mathit{reflect}}$ and $\mathbf{x}_{\mathit{expand}}$ and the iteration is terminated.

Contract: If $f(\mathbf{x}_{reflect}) \geq f(\mathbf{x}_N)$, i.e. if the reflected point would still be the worst if it replaced the worst point so far, then a contraction is attempted. Depending on whether the reflected point is better or worse than the worst point so far, two types of contraction are possible.

Outside: If $f(\mathbf{x}_{reflect}) \leq f(\mathbf{x}_{N+1})$, then an outside contraction point

$$\mathbf{x}_{contract} = \langle \mathbf{x} \rangle + \beta(\mathbf{x}_{reflect} - \langle \mathbf{x} \rangle)$$

is computed and its objective function value is evaluated.

Inside: If $f(\mathbf{x}_{reflect}) > f(\mathbf{x}_{N+1})$, then an inside contraction point

$$\mathbf{x}_{contract} = \langle \mathbf{x} \rangle + \beta(\mathbf{x}_{N+1} - \langle \mathbf{x} \rangle)$$

is computed and its objective function value is evaluated.

In either case, if $f(\mathbf{x}_{contract}) \leq \min(f(\mathbf{x}_{N+1}), f(\mathbf{x}_{reflect}))$, then the worst point $\mathbf{x}_{N+1}$ is replaced by $\mathbf{x}_{contract}$ and the iteration is terminated.

Shrink: If all of the above have failed to generate a point that is better than the second worst, then all vertices $\mathbf{x}_i$ but the best are replaced by new points

$$\mathbf{x}'_i = \mathbf{x}_1 + \delta(\mathbf{x}_i - \mathbf{x}_1), \qquad i = 2, \ldots, N+1,$$

and objective function values of the new points are evaluated. The iteration is terminated.

According to Nelder and Mead, the purpose of these operations is that "the simplex adapts itself to the local landscape, elongating down inclined planes, changing direction on encountering a valley at an angle, and contracting in the neighborhood of a minimum". Clearly, depending on the quality of the new points that are generated, the method requires either 1, 2, or $N+2$ objective function evaluations per time step. The almost universally recommended values for the parameters are $\alpha = 1$, $\beta = \delta = 1/2$, and $\gamma = 2$. Figure 6.1 shows the effects of reflection, expansion, contraction, and shrinkage for $N = 2$. Note that as is common practice today, the expansion step slightly deviates from the original prescription by Nelder and Mead in that comparison of the objective function value of the expansion point is with $f(\mathbf{x}_{reflect})$ rather than with $f(\mathbf{x}_{N+1})$. A good description of the method along with a discussion of its properties has been published by Wright [92].

According to Barton and Ivey [16], in 1996 the simplex method of Nelder and Mead was the most popular direct search strategy based on published applications. In 1995, Elster and Neumaier [35] asserted that the simplex method was also the usually recommended and the most frequently used method for noisy function optimization. Simplex code can be found in *Numerical Recipes* [63] and is part of the *Matlab* optimization toolbox.

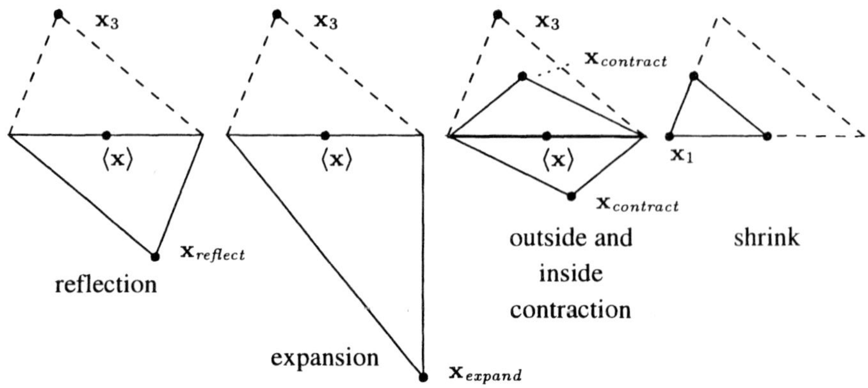

Figure 6.1. Nelder-Mead simplices after a reflection, an expansion, an outside contraction, an inside contraction and a shrink step. The original simplex is shown with dashed lines, the transformed simplex with solid lines.

In spite of its widespread use, it is well known that the performance of the simplex method of Nelder and Mead frequently is insufficient. A reason for the bad performance of the method has been identified in the tendency of the simplices to collapse into a subspace of the search space or to become extremely elongated and distorted in shape even if the local structure of the objective function does not demand that. In an interesting paper, McKinnon [56] constructed a two-dimensional, strictly convex objective function that has continuous second derivatives where the simplex method converges to a nonoptimal point. The method repeatedly applies the inside contraction step with the best vertex remaining fixed. As a consequence, the simplices tend to a straight line which is orthogonal to the steepest descent direction. Moreover, Torczon [86] has shown experimentally that the algorithm of Nelder and Mead fails even on the sphere unless the dimensionality N of the search space is very small, and that the presence of noise even worsens the tendency of the method to stagnate at nonoptimal points.

A number of fixes to the simplex algorithm have been suggested. Most of those fixes rely on restart techniques if the simplex becomes too degenerate or — in the presence of noise — on resampling strategies. Restart strategies have been suggested by Humphrey and Wilson [48] and by Kelley [49]. In the latter reference, a test for detecting stagnation of the search is proposed and it is suggested that a restart with a smaller simplex with an orientation that is determined by a descent direction from the current best point be initiated in that case. Barton and Ivey [16] reported that the performance of the algorithm in the presence of noise can be improved by demanding that the best point in the simplex be reevaluated in each shrink step and by choosing $\delta = 0.9$ rather than $\delta = 0.5$.

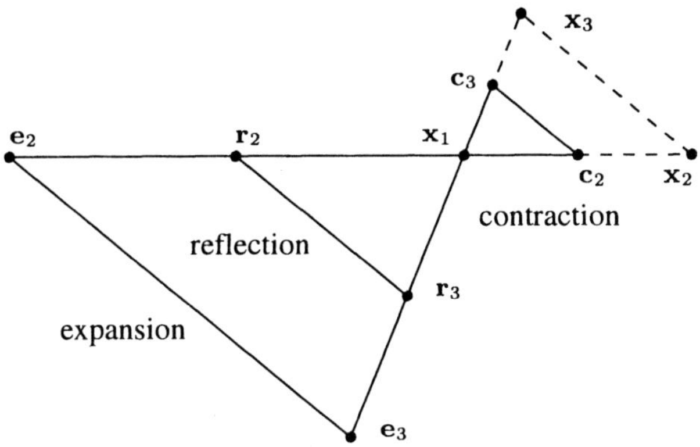

Figure 6.2. Multi-directional search reflection, expansion, and contraction. The original simplex is shown with a dashed line.

1.3 Multi-Directional Search

The multi-directional search algorithm of Torczon [86] is a simplex-based strategy that attempts to overcome the shortcomings of the algorithm of Nelder and Mead. A primary motivation for the new method was the desire for efficiency in a parallel computing environment. An empirical comparison of multi-directional search with the simplex method of Nelder and Mead led Torczon to suggest that "Our preliminary performance results give us reason to believe that the multi-directional search algorithm may prove to be most useful when the function evaluations are subject to error". A related method has recently been suggested by Anderson and Ferris [2].

Realizing that degenerate simplices are a frequent source of failure of the simplex method of Nelder and Mead, Torczon insisted that for the multi-directional search method the shape of the simplices does not change but that merely their size varies. In contrast to the Nelder-Mead method, not single vertices but the entire simplex is reflected, expanded, and contracted in one time step. An iteration succeeds when it finds a point of strict improvement over the *best* vertex, in contrast to the much weaker condition in a Nelder-Mead iteration of finding a strict improvement compared to the worst point. Again, we assume that the simplex vertices have been ordered such that $\mathbf{x}_1$ is the vertex with the best objective function value. The sequence of operations constituting a step of the multi-directional search method is as follows:

Reflect: The reflected vertices

$$\mathbf{r}_i = \mathbf{x}_1 + \alpha(\mathbf{x}_1 - \mathbf{x}_i), \qquad i = 2, \ldots, N+1,$$

are computed and their objective function values are evaluated. In case $\min_i\{f(\mathbf{r}_i)\} < f(\mathbf{x}_1)$, then an expansion is attempted; otherwise, a contraction is performed.

Expand: The expanded vertices

$$\mathbf{e}_i = \mathbf{x}_1 + \gamma(\mathbf{x}_1 - \mathbf{x}_i), \qquad i = 2, \ldots, N+1,$$

are computed and their objective function values are evaluated. If the best vertex so far is improved on, i.e. if $\min_i\{f(\mathbf{e}_i)\} < \min_i\{f(\mathbf{r}_i)\}$, then the expanded vertices $\mathbf{e}_i$ replace the vertices $\mathbf{x}_i$. Otherwise, the reflected vertices $\mathbf{r}_i$ replace the vertices $\mathbf{x}_i$. In either case, the iteration is terminated.

Contract: The contracted vertices

$$\mathbf{c}_i = \mathbf{x}_1 - \beta(\mathbf{x}_1 - \mathbf{x}_i), \qquad i = 2, \ldots, N+1,$$

are computed and their objective function values are evaluated. The contracted vertices $\mathbf{c}_i$ replace the original vertices $\mathbf{x}_i$, and the iteration is terminated.

Recommended values of the parameters are $\alpha = 1$, $\beta = 1/2$, and $\gamma = 2$. Figure 6.2 shows the effects of reflection, expansion, and contraction for $N = 2$. Clearly, an iteration of the multi-directional search requires $2N$ evaluations of the objective function.

1.4 Implicit Filtering

In contrast to the search methods introduced so far, implicit filtering as devised by Gilmore and Kelley [38, 50] relies on explicitly approximating the local gradient of the objective function by means of finite differencing. A brief summary of the algorithm due to Kelley [50] states that "In its simplest unconstrained form, implicit filtering is the steepest descent algorithm with difference gradients, where the difference increment varies as the iteration progresses. Because the gradient is only an approximation, the computed steepest descent direction may fail to be a descent direction and the line search may fail. In this event, the difference increment is reduced." The name "implicit filtering" has been chosen because the method uses the differencing to "step over" the noise at varying levels of resolution, hence implicitly filtering the objective. It is worth noting that Kelley [50] has enhanced the basic algorithm to be described below by a quasi-Newton component that attempts to accumulate second-order information on the objective function in the course of the search. As the sphere is the only objective function we attempt to minimize, we do not consider this extension here. Kelley [50] also provides pointers to optimization problems that implicit filtering has been applied to.

The state of the implicit filtering algorithm at time t is described by a base point $\mathbf{x}$ and a difference increment h. Writing $\mathbf{e}_i$ for the ith unit vector, a central finite difference gradient $\nabla_h f(\mathbf{x})$ with ith component

$$(\nabla_h f(\mathbf{x}))_i = \frac{f(\mathbf{x} + h\mathbf{e}_i) - f(\mathbf{x} - h\mathbf{e}_i)}{2h}$$

is computed by taking steps of length h in both the positive and negative directions parallel to the axes of the coordinate system. Clearly, this step involves $2N$ evaluations of the objective function. Instead of using central differences, forward differences could be employed; however, Kelley [50] states that the performance of implicit filtering with a central difference gradient is far superior to that with forward difference gradients.

Subsequently, a line search in the negative direction of the approximate gradient thus obtained is carried out. In implicit filtering, the algorithm of Armijo [4] is the usual choice for a line search method. That is, starting with $\lambda = \lambda_0$, it is tested whether the condition

$$f(\mathbf{x}) - f(\mathbf{x} - \lambda \nabla_h f(\mathbf{x})) \geq \frac{1}{2}\lambda \|\nabla_h f(\mathbf{x})\|^2$$

holds. If it does hold, then the base point is replaced by $\mathbf{x} - \lambda \nabla_h f(\mathbf{x})$ and the implicit filtering algorithm proceeds to the next iteration. Otherwise, λ is halved and the condition is tested again. If the value of λ has been halved i_{max} times, then the line search is aborted, the difference increment h is halved, and a new iteration of the algorithm is started with the base point left unchanged. Clearly, the line search maximally requires a number of objective function evaluations that depends on i_{max}, but not on the dimensionality N of the search space if i_{max} is chosen independent of N.

1.5 Simultaneous Perturbation Stochastic Approximation

The simultaneous perturbation stochastic approximation (SPSA) algorithm of Spall [78, 79, 80] attempts to improve on classical stochastic approximation algorithms by not obtaining a regular finite difference approximation to the gradient in every iteration — an operation that requires a large number of objective function evaluations in high-dimensional search spaces — but to obtain an approximation by means of a procedure that requires only two objective function evaluations regardless of the dimensionality of the search space. Spall [80] contended that "SPSA [...], relative to other methods, is most effective when the loss function measurements include added noise" and provided pointers to a number of noisy optimization problems that the SPSA algorithm has been applied to. Spall [81] also introduced an extension of the basic SPSA algorithm that attempts to gather second-order information on the objective function with a small fixed number of additional fitness function evaluations. As the corresponding extension to implicit filtering, the second-order component of SPSA is of little interest for the purpose of minimizing the sphere and is omitted in what follows.

To obtain an approximation to the gradient of the objective function, a random vector $\mathbf{z} \in I\!R^N$ with components that are drawn independently and with equal probability from $\{\pm 1\}$ is generated. Spall [78] has shown that the vector with

ith component

$$(\mathbf{d}_h(\mathbf{x}))_i = \frac{f(\mathbf{x} + h\mathbf{z}) - f(\mathbf{x} - h\mathbf{z})}{2hz_i} \tag{6.1}$$

is an almost unbiased estimator for the gradient of the objective function provided that h is sufficiently small. Note that two objective function evaluations are sufficient to obtain $\mathbf{d}_h(\mathbf{x})$ while $2N$ evaluations are required to obtain $\nabla_h f(\mathbf{x})$.

As usual in stochastic approximation, the original SPSA algorithm uses fixed schedules for reducing the size of thc difference increments h and the lengths of the steps made in the direction of the approximate gradients. The commonly used schedules are much too slow for our purposes. Therefore, we test the simultaneous perturbation gradient approximation $\mathbf{d}_h(\mathbf{x})$ by substituting it for the gradient estimate $\nabla_h f(\mathbf{x})$ in the implicit filtering algorithm. The Armijo line search component and the scheme for updating the difference increment h remain unchanged.

1.6 Evolutionary Gradient Search

The evolutionary gradient search algorithm of Salomon [74] is a search method that employs yet another, "evolutionary inspired" estimation of the local gradient of the objective function. It also differs from the two previously introduced gradient-based algorithms in that it employs a form of adaptation of the step length that is a mixture of a line search and ideas from mutative self-adaptation methods in evolution strategies.

The state of the evolutionary gradient search algorithm at time t is described by a base point $\mathbf{x}$ and a step length h. In every iteration, λ offspring candidate solutions $\mathbf{x} + h\mathbf{z}_i$, $i = 1, \ldots, \lambda$, are generated, where the $\mathbf{z}_i$ are random vectors with independent components drawn from a standardized normal distribution. Rather than performing selection the way an evolution strategy does, the evolutionary gradient search method computes

$$\mathbf{d}_h(\mathbf{x}) = \frac{1}{\lambda}\sum_{i=1}^{\lambda} \left[f(\mathbf{x} + h\mathbf{z}_i) - f(\mathbf{x})\right] \mathbf{z}_i \tag{6.2}$$

as an estimation of the local gradient of the function. The motivation for this step is the wish to not discard the information carried by the offspring that are rejected, but to interpret an offspring candidate solution with a negative fitness advantage over the base point as evidence that a step should be taken in the opposite direction. The method then proceeds by taking two test steps from the base point in the negative direction of $\mathbf{d}_h(\mathbf{x})$. One test step has length $h\sqrt{N}\zeta$, the other one has length $h\sqrt{N}/\zeta$, where usually $\zeta = 1.8$. The base point is then updated by performing the test step with the higher (measured) fitness advantage, and the step length h is multiplied by ζ if the longer of the two test steps was more successful and it is divided by ζ if the shorter of the two test steps prevailed. Clearly, an iteration of the evolutionary gradient search procedure requires $\lambda + 2$ evaluations of the objective function.

It is interesting to compare the direction given by the negative of Equation (6.2) in which the evolutionary gradient search method proceeds with that of other strategies. In the absence of noise, Salomon has shown that for small h and sufficiently large λ, the direction given by Equation (6.2) agrees closely with the gradient direction at location $\mathbf{x}$. For $\lambda = 1$ on the other hand, the difference between the evolutionary gradient search direction and the SPSA direction given by Equation (6.1) resembles the difference between a forward difference approximation of the gradient and a central difference approximation thereof. Note however that the mutation vectors are drawn from different probability distributions. Comparing the evolutionary gradient search method with a $(\mu/\mu, \lambda)$-ES, it is striking that both strategies employ a weighted averaging of mutation vectors. While the $(\mu/\mu, \lambda)$-ES assigns equal weights to the μ best of the λ offspring and zero weights to the remaining ones, the evolutionary gradient search method assigns weights that are proportional to the fitness advantages of all offspring, with offspring with a negative fitness advantage receiving negative weights. While the evolutionary gradient search method has not yet been studied in great detail, it seems conceivable that the genetic repair effect may be present in evolutionary gradient searches.

There are at least two extensions to the basic evolutionary gradient search algorithm. Salomon [74] proposed the use of a momentum term as frequently employed in the neural network literature that provides a memory of the most recently taken steps. In another paper, Salomon [75] also introduced a heuristic to handle N step sizes rather than just a single one in an attempt to improve the performance of the method on badly scaled objective functions. Neither of the two extensions is of use for the minimization of the sphere.

2. The Competition

All of the strategies thus introduced have been tested extensively on noisy spheres of several search space dimensionalities. Section 2.1 provides a detailed description of the computer experiments that have been conducted. Section 2.2 describes and discusses the results that have been obtained.

2.1 Experimental Setup

The fitness environment in which we have tested the search methods differs from the spherical environment introduced in Section 3 of Chapter 2 in the single point that we drop the assumption that there is a common noise strength for all offspring candidate solutions generated in a single time step that is determined by the centroid of the parental location rather than by that of the offspring

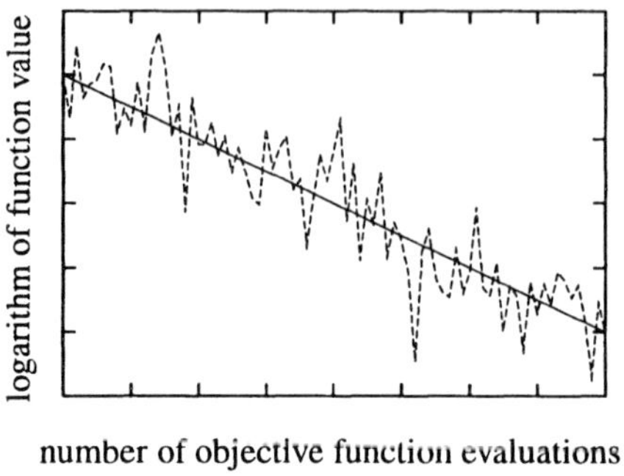

Figure 6.3. Typical convergence behavior of direct search strategies on the noisy sphere. The dashed line shows the logarithm of the strategy's function value over the number of objective function evaluations. The efficiency of the search strategy is determined by the negative of the slope of the solid line.

being evaluated. The assumption was made so as to facilitate the mathematical analyses in Chapters 3 through 5. In search spaces of high dimensionality, for some strategies such as the $(1+1)$-ES and the (μ, λ)-ES in which the distance between parents and offspring ideally is relatively short, the difference in performance resulting from dropping the assumption is rather minor. However, in low-dimensional search spaces and for strategies that spread their offspring much wider, the difference can be considerable. We feel that it is necessary to drop the assumption as search strategies such as implicit filtering that employ different step lengths for their testing and operating steps could effectively eliminate the noise merely by increasing the length of their testing steps.

In addition to modifying the fitness environment, we also modify the definition of the efficiency of a strategy originally introduced in Section 4 of Chapter 2. We shall refer to the ideal objective function value of the centroid of all candidate solutions at time t as the strategy's function value of that time step. We could as well consider the ideal objective function value of the best candidate solution or the average of the ideal function values of all candidate solutions instead. The particular choice is without relevance for what follows. Due to the scale invariance of the sphere, when plotting the logarithm of the strategy's function value over the number of objective function evaluations, the resulting graph typically is linear with some superimposed fluctuations as illustrated in Figure 6.3. In line with the definition of the efficiency in previous chapters, we consider the negative of the slope of the regression line as a measure for the efficiency of the strategy.

So as to determine the slope of the line, we need to find the average change in the strategy's logarithmic function values from one time step to the next as well as the average number of objective function evaluations per time step. For the purpose of the former, let us denote the strategy's function value at time t by R^2 and the function value of the succeeding time step by r^2. We redefine the strategy's normalized quality gain on the sphere as the expected value of

$$\Delta^*_{sphere} = -\frac{N}{2} \log\left(\frac{r^2}{R^2}\right). \tag{6.3}$$

Note that if $r^2 \approx R^2$, i.e. if the strategy's function value changes only slowly as it typically does for large N, then the logarithm can be expanded into a Taylor series around unity that is cut off after the linear term and the normalized quality gain defined here formally agrees with the normalized quality gain defined in Chapter 2. Only for small values of N do the two quantities differ markedly.

Thus, to determine the efficiency of a strategy, we run it for a number of time steps t, $0 \leq t < t_{end}$. We determine the average normalized quality gain and the average number of objective function evaluations in a time interval $t_{start} \leq t < t_{end}$. Note that t_{start} and $t_{end} - t_{start}$ have to be large enough for initialization effects to fade and for fluctuations to average out, respectively. In our computer experiments, we use $t_{start} = 2,000$ and $t_{end} - t_{start} = 40,000$. We use the quotient of the two averages thus obtained as an estimate for the efficiency η of the strategy.

2.2 Experimental Results

Clearly, the strategies outlined in Section 1 employ quite different approaches to obtaining descent directions. While some strategies place new candidate solutions deterministically, others generate them randomly. Further differences consist in the number of new candidate solutions that are generated per time step. Some strategies, such as the simultaneous perturbation approach, conduct only a small, fixed number of objective function evaluations before taking a step. Other strategies, such as implicit filtering or the multi-directional search method, utilize a number of objective function evaluations that is proportional to the search space dimensionality N. The two evolutionary-inspired approaches leave the number of objective function evaluations per time step as a parameter that can be chosen freely. In general, improved accuracy of the approximation to the gradient direction can be obtained at the price of increased computational costs per time step.

However, not only the direction in which a step is taken, but also the step length control mechanism is of vital importance for the performance of a search strategy on the sphere. Not only does it have a quantitative influence on the efficiency, but the step length adaptation component also determines in what range of noise strengths the convergence behavior described in Section 2.1 and illustrated in Figure 6.3 is attained. For almost all of the search strategies considered, there exists a noise strength above which convergence becomes unreliable. In most cases, this effect shows in a rapid decrease of step lengths and leads to stagnation of the search process. We therefore monitor the step lengths of the strategies so as to detect failure. Typically, such failure can as well easily be recognized visually in the plots of the logarithmic function values that become horizontal at some point. Those runs in which the kind of convergence behavior illustrated in Figure 6.3 was not achieved are excluded from the results

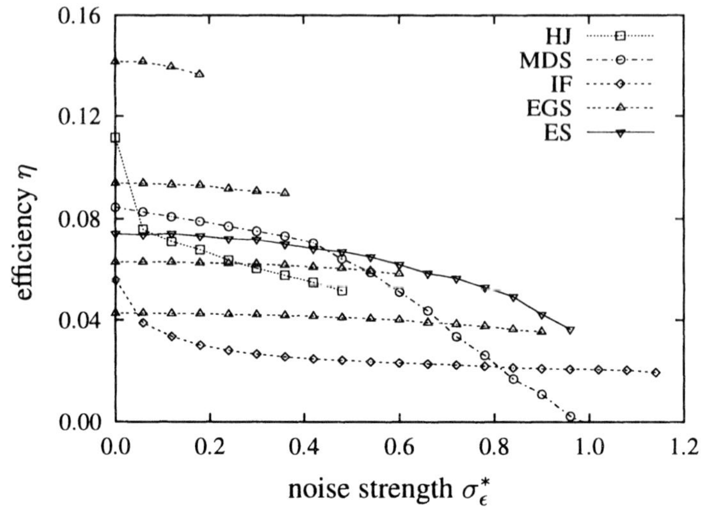

Figure 6.4. Efficiency η of search strategies as functions of normalized noise strength σ_ϵ^* for search space dimensionality $N = 4$. The curves represent results for direct pattern search (HJ), multi-directional search (MDS), implicit filtering (IF), evolutionary gradient search (EGS) with, from top to bottom, $\lambda = 4, 10, 20,$ and 40, and a $(2/2, 6)$-ES with cumulative mutation strength adaptation (ES).

presented below. The corresponding curves in the figures thus end abruptly at some point.

Finally, it needs to be noted that for those algorithms in which search points can survive for an unlimited number of time steps, periodic reevaluation of the fitness function values of the surviving search points is required in the presence of noise so as to prevent stagnation due to accumulated overvaluation — a phenomenon studied in Section 3.3 of Chapter 3 for the case of the $(1+1)$-ES. So as to have consistent behavior across all of our measurements, we apply reevaluation strategies even in the absence of noise. The results of measuring the performance of the search strategies on the noisy sphere with search space dimensionalities $N = 4$, 40, and 400 are shown in Figures 6.4, 6.5, and 6.6, respectively, and are described in detail in what follows.

Direct Pattern Search In our implementation of the direct pattern search algorithm of Hooke and Jeeves, the objective function value of the base point is reevaluated in every time step. Failure of achieving the type of convergence behavior described in Section 2.1 is marked by a rapid decrease in the step length. It can be seen that the efficiency of the method is quite good in the absence of noise, but rather rapidly declines if there is noise present. This is especially true in the higher-dimensional search spaces where the efficiency is competitive only for zero noise strength.

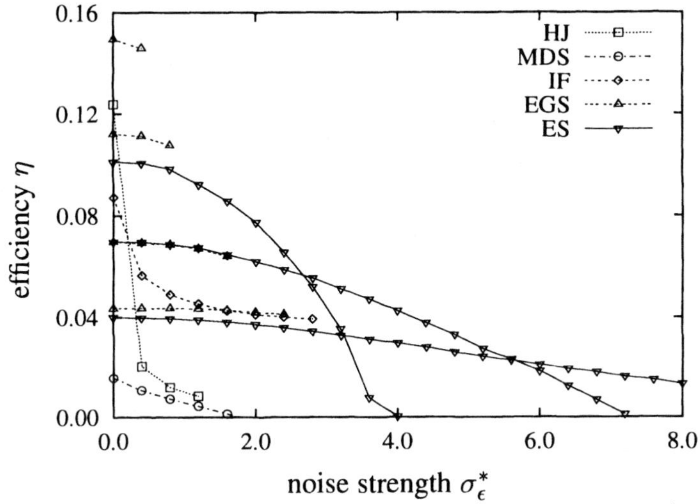

Figure 6.5. Efficiency η of search strategies as functions of normalized noise strength σ_ϵ^* for search space dimensionality $N = 40$. The curves represent results for direct pattern search (HJ), multi-directional search (MDS), implicit filtering (IF), evolutionary gradient search (EGS) with, from top to bottom, $\lambda = 20, 40, 100$, and 200, and, from top-left to bottom-right, a $(3/3, 10)$-ES, a $(6/6, 20)$-ES, and a $(12/12, 40)$-ES with cumulative mutation strength adaptation (ES).

Simplex Method Our observations of the performance of the simplex method of Nelder and Mead on the noisy sphere agree with those of Torczon [86]. For $N = 4$ and zero noise strength, we have obtained an efficiency of about 0.26, thus exceeding that of all other search strategies we have tested but implicit filtering. However, for nonzero noise strength or for $N = 40$ or even $N = 400$, not a single run of the strategy resulted in the type of convergence behavior described in Section 2.1. We have tested a number of reevaluation strategies and have followed the recommendations of Barton and Ivey [16], but to no discernible effect. There is little use in averaging over multiple samples as almost any nonzero level of noise can lead to stagnation. We have not employed any restart strategies as the breakdowns are so frequent that the behavior of the algorithm would be determined by the restart strategy rather than by the simplex search.

Multi-Directional Search In our implementation of Torczon's multi-directional search method, we employ a regular simplex, i.e. one for which all edges have the same length. In order to achieve the type of convergence behavior described in Section 2.1, it is necessary to reevaluate the objective function value of the best vertex in every time step. In contrast to most of the other search strategies, the multi-directional search method never stagnates but rather diverges if the noise strength is too high.

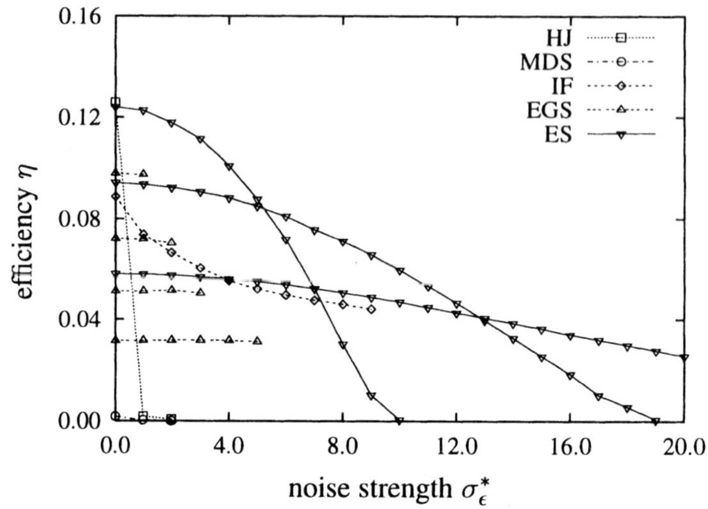

Figure 6.6. Efficiency η of search strategies as functions of normalized noise strength σ_ϵ^* for search space dimensionality $N = 400$. The curves represent results for direct pattern search (HJ), multi-directional search (MDS), implicit filtering (IF), evolutionary gradient search (EGS) with, from top to bottom, $\lambda = 100$, 200, 400, and 1000, and, from top-left to bottom-right, a $(6/6, 20)$-ES, a $(12/12, 40)$-ES, and a $(24/24, 80)$-ES with cumulative mutation strength adaptation (ES).

It can be seen that overall, the efficiency of the strategy on the noisy sphere is satisfactory only for $N = 4$. In contrast to most of the other methods considered, the efficiency markedly declines with increasing search space dimensionality and is virtually zero for $N = 400$ even in the absence of noise. While in the absence of noise and for low noise strengths the strategy's suitability for implementation on a parallel computer may be an asset, parallelization would merely lead to faster divergence in the range of noise strengths in which the efficiency is negative.

Implicit Filtering In the absence of noise, implicit filtering converges much faster than any of the other methods as the exact gradient direction is obtained. Whether the exact optimizer is attained in a single time step thus only depends on the outcome of the line search. Rather than measuring the performance of the implicit filtering algorithm for zero noise strength, we have used the very low noise strength $\sigma_\epsilon^* = 0.001$ instead.

It can frequently be read that finite-difference gradient-based methods exhibit poor performance in the presence of noise. This is true if the difference increments are so small that the differences in function values are hidden by the noise. However, implicit filtering uses difference increments that are large enough to afford quite good performance. Only for relatively high noise strengths does the strategy fail to achieve the kind of convergence behavior described above. Closer investigation reveals that if h is too small, then the gradient approximation is so unreliable that the line searches fail frequently and h is decreased further, leading to stagnation of the search.

While the parameter λ_0 that determines the maximum step length and that was set to unity in our experiments is relatively uncritical as long as it is chosen large enough, the maximum number of iterations per line search i_{max} does have a decisive influence on the performance of the algorithm. In our experiments, we have used $i_{max} = 8$. In general, the size h of the difference increments is above its optimal value. Decreasing i_{max} leads to it being decreased faster and therefore to improved efficiency, but at the price of decreased stability.

Simultaneous Perturbation Stochastic Approximation The attempt to use the SPSA gradient approximation instead of computing the central difference gradient in the implicit filtering scheme has failed altogether. Even for zero noise strength is the gradient approximation so inexact that the line search frequently fails. As a consequence, the size of the difference increment h quickly tends to zero and the search stagnates. Note however that the original SPSA algorithm requires fixed schedules for both the difference increments and the step lengths, and that a better adaptation strategy for the difference increments and the step lengths might make the SPSA method of approximating gradients useful in the current context.

Evolutionary Gradient Search Quite interestingly, the evolutionary gradient search method proves to be one of the most efficient strategies apart from implicit filtering in the absence of noise. Without noise, the number of candidate solutions generated per time step ideally is rather small. In the presence of noise however, much larger values of λ are required for the strategy to perform satisfactorily, and the efficiency suffers. If λ is chosen too small, then the step length essentially performs a random walk. The dependence of the optimal number of candidate solutions per time step on the noise strength and on the search space dimensionality remains to be studied.

Evolution Strategy The evolution strategy with cumulative mutation strength adaptation clearly is that strategy in the lineup which is the most robust with regard to the effects of noise. For $N = 40$ and especially for $N = 400$, it is the only search strategy that converges reliably for higher levels of noise. As for choosing appropriate population sizes, Figure 5.13 provides useful orientation. It can be seen from Figures 6.5 and 6.6 that overall the choice is not very critical, and that satisfactory performance can usually be achieved for a range of population size parameter settings.

To conclude, a number of interesting things can be learned from a comparison of the performance of the search strategies. First, we were surprised to see how little the efficiencies of the various strategies differ in the absence of noise. With the exceptions of implicit filtering which has the benefit of being able to work with exact gradients, the simplex method of Nelder and Mead that has severe performance deficits, and the multi-directional search method which does not scale well with the dimensionality of the search space, the efficiencies of all methods we have considered are of the same order of magnitude, irrespective of the number and location of candidate solutions generated per time step. Second, it has been seen that the usual road to failure for most search strategies in the presence of noise consists in the unwarranted decrease of difference increments or step lengths. For many step length adaptation mechanisms, small step lengths tend to lead to a further reduction of step lengths and therefore to stagnation. In the presence of noise, extra care must be taken to avoid getting caught in such a downward spiral. Finally, the evolution strategy with cumulative mutation strength adaptation is the only one of the strategies we have considered that uses information accumulated over a number of time steps for the adaptation of its step length, and at the same time it is that strategy which performs best in the presence of noise. It seems reasonable to conclude that it is the averaging effect that is inherent in the accumulation of the progress vectors makes cumulative mutation strength adaptation relatively robust with regard to the effects of noise.

Chapter 7

CONCLUSIONS

In this book, we have studied the effects that noise has on the local performance of several variants of the $(\mu/\rho \overset{+}{,} \lambda)$-ES. Our conviction that certain behaviors of evolution strategies can best be understood in the most simple environments in which they can be observed led us to the definition of the linear and of the spherical fitness environments in Section 3 of Chapter 2. Inherent to both fitness environments are symmetries that result in invariance properties that simplify substantially the analysis of the behavior of the strategies. After initialization effects have faded, the behavior of the strategies can be described by time-invariant probability distributions. Most of the analyses presented in this work followed the same basic pattern: identify the variables determining the strategies' state, determine the effects that variation and selection have on them, and infer information on those variables by means of invariance considerations. As the variables we sought to learn about are random variables, those steps required the handling of probability distributions of usually unknown shape, and frequently we had to resort to expanding the distributions in terms of derivatives of a normal distribution and to hope that they are well characterized by a number of lower-order moments. That hope and various modeling assumptions and simplifications made in the calculations made it necessary to verify the accuracy of our results by comparing them with results of computer experiments. We took good agreement between results obtained analytically and results observed in experiments as evidence that what we have learned from the simple models is useful for the understanding of the more complex dynamical systems. In this chapter, we conclude with a summary of the most important insights we have gained and with some suggestions for future research.

In Chapter 3, we have analyzed the performance of the $(1 + 1)$-ES on the noisy sphere in the limit of infinite search space dimensionality. Good agreement of the results with empirical measurements in a search space of dimensionality $N = 40$ has been demonstrated. The analysis differs from that presented by Beyer [18] in that we did not require that the fitness of surviving candidate solutions be reevaluated in every time step. Crucial differences in performance between the two types of strategy could be observed. We identified systematic overvaluation of the parental fitness as the reason for those differences, and we determined estimates of mean and variance of the degree of normalized overvaluation under the assumptions that its distribution is normal and that constancy of the normalized mutation strength is assured. We found that while overvaluation of the parental fitness reduces success probabilities and can lead to long periods of stagnation, its influence on the quality gain of the $(1+1)$-ES can be overall beneficial. Overvaluation prevents the strategy from diverging due to noisy fitness measurements. This effect has been used by Markon et al. [55] who employed explicit thresholds rather than the implicit overvaluation of the parental fitness to achieve improved performance of a $(1+1)$-ES in the presence of noise. Subsequently, we studied the potential of averaging over a number of independent fitness measurements as a way to achieve improved performance in the presence of noise, and we found that such resampling is beneficial for the efficiency of the strategy above a certain, relatively low noise strength. Finally, we investigated the effects that noise has on a simple, success probability-based mutation strength adaptation rule. We found that overvaluation can render the rule useless above a certain noise strength, and that simply tweaking the target success probability cannot solve the problem.

In Chapter 4, the influence of distributed populations on the performance of evolution strategies has been studied. In particular, the behavior of the (μ, λ)-ES in the linear fitness environment was analyzed using a moment-based approach for describing the population of candidate solutions. The effort involved in the analysis exceeded that of all other parts of this work. In particular, in Appendix D, we introduced noisy order statistics as a generalization of ordinary order statistics, and we derived joint probability density functions of noisy order statistics and computed expected moments of truncated samples, where the truncation criterion had to rely on noisy fitness information. The results of the analysis that included a number of simplifying modeling assumptions proved to yield quite good estimates for the lower-order central moments of the population as well as for the strategy's quality gain. Based on those results, we identified the variance of the population of candidate solutions as a quantity of great importance for the understanding of the (μ, λ)-ES. We saw that it contributes to the "signal strength" of the selection process, and that greater variances improve the signal-to-noise ratio that the strategy operates under. However, as greater variances can be achieved only by increasing the proportion of candidate solutions that are retained and thus at the price of reduced selection pressure, there exists an optimal truncation ratio that depends on the population size as well as on the noise strength. We have seen that the optimal truncation ratio increases as the noise strength increases, but that attaining it exactly becomes less significant as the dependence of the quality gain on it becomes weaker.

Subsequently, we have made plausible that the results obtained in the linear environment do bear relevance to the sphere, and we have shown empirically that for sufficiently high search space dimensionality, the quality gain of the (μ, λ)-ES on the sphere can be reduced to that in the linear environment. Determining optimal population sizes on the sphere we found that the optimal number of offspring per time step is much lower for the (μ, λ)-ES than for the $(1, \lambda)$-ES, and that above a certain noise strength substantial performance gains can be achieved by retaining more than the (seemingly) best candidate solution.

No attempt has been made to analyze the performance of mutation strength adaptation mechanisms for the (μ, λ)-ES. An investigation of the behavior of such mechanisms remains as a challenge for the future. Moreover, it would be desirable to find a more simple law than that derived in Chapter 4 that at least qualitatively describes in a more intuitive fashion how the quality gain of the (μ, λ)-ES in the linear environment scales with the population size parameters μ and λ. An empirical observation with regard to the residual location error of the (μ, λ)-ES on the noisy sphere made by Beyer [26] suggests that such a simple relationship may exist. Whether it can be found by simplifying the result derived here or whether an entirely different approach to analyzing the performance of the strategy is required is an open question.

In Chapter 5, the effects of global intermediate recombination in the presence of noise have been studied. A simple quality gain formula for the $(\mu/\mu, \lambda)$-ES in the spherical environment in the limit of infinite search space dimensionality has been derived that neatly generalizes several previously obtained results and that confirms a number of previously made hypotheses and observations. It has been seen that the $(\mu/\mu, \lambda)$-ES can partially compensate for a lack of reliable information by using the noisy information provided by a larger number of parents than it would optimally use in the absence of noise. The simple quality gain law is also valuable for explaining the significant performance advantages that result from global intermediate recombination. As found by Rechenberg [67] and Beyer [21] in the absence of noise, genetic repair, i.e. the reduction in length of the "harmful" components of mutations as a result of the averaging process inherent in intermediate recombination, makes it possible to operate the $(\mu/\mu, \lambda)$-ES at much higher mutation strengths than those strategies that do not use recombination. In the presence of noise, genetic repair has the added benefit that higher mutation strengths lead to a reduced noise-to-signal ratio. As a consequence, it has been seen that for the $(\mu/\mu, \lambda)$-ES — and in contrast to the $(1, \lambda)$-ES — increasing the population size and the number of offspring candidate solutions generated per time step is preferable to averaging over multiple samples. We have argued that a strategy for which this is not true cannot be considered to be able to cope with noise as explicitly removing the noise is preferable to letting the strategy deal with it.

Unfortunately though, the accuracy of the simple quality gain law in finite-dimensional search spaces is not very good for all but the smallest population sizes. While it can be used to infer that the efficiency of the $(\mu/\mu, \lambda)$-ES can be increased up to its optimal value in the absence of noise by increasing the population size and the number of offspring candidate solutions generated per time step, this is not what can be observed in practice. We have therefore derived an improved approximation to the quality gain of the $(\mu/\mu, \lambda)$-ES on the finite-dimensional sphere that lacks the conciseness of the previously derived formula, but that is useful for studying numerically the scaling properties of the strategy. It has been seen that in finite-dimensional search spaces, genetic repair cannot be as effective as the results obtained for infinite search space dimensionality suggest. It is not useful to increase the number of candidate solutions generated per time step beyond a certain number that increases with increasing noise strength and with increasing search space dimensionality. It has been seen that overall, that number is rather moderate compared with the search space dimensionality, and that the choice of the number of candidate solutions to generate per time step becomes less critical as the search space dimensionality increases. It has also been seen that the maximal efficiency of the (μ, λ)-ES and, except for very low noise strength, that of the $(1+1)$-ES can be far exceeded by the $(\mu/\mu, \lambda)$-ES.

In Section 3 of Chapter 5, the behavior of cumulative mutation strength adaptation of a single mutation strength on the noisy sphere has been investigated. We have seen that except for zero noise strength, the target mutation strength, i.e. the mutation strength that the mutation strength adaptation scheme seeks to attain, is below the mutation strength that optimizes the performance of the strategy. However, we have also seen that dynamical effects play a role as well, and that the average mutation strength that is realized by the $(\mu/\mu, \lambda)$-ES with cumulative mutation strength adaptation may even exceed the optimal value. By studying the logarithmic adaptation response, we have seen that in the presence of noise, if the current mutation strength is far too low, the tendency towards higher mutation strengths is very weak. This is plausible as the goal of cumulative mutation strength adaptation is to achieve that consecutive steps are uncorrelated. If the steps are so small that the information they carry is hidden by noise, then this goal is achieved and the mutation strength adaptation scheme fails to recognize the need to increase the step length. It may thus be useful to either start with a sufficiently high mutation strength or to incorporate a tendency to higher mutation strengths into the algorithm if it is known that fitness evaluations are subject to noise. Finally, based on the improved approximation of the quality gain from Section 2 of Chapter 5, we have obtained estimates for optimal population sizes and the corresponding efficiencies. Optimal population sizes have been seen to be even smaller than they are for optimally adapted mutation strength. The efficiencies that are obtained are quite good, and the performance of cumulative mutation strength adaptation has been seen to be solid provided that population sizes are appropriately chosen. It has to be noted however that even though the accuracy of the estimates for the quality gain on the noisy sphere of the $(\mu/\mu, \lambda)$-ES with cumulative mutation strength adaptation is quite good, the results have been obtained under the assumption that fluctuations can be ignored. A goal of future research will be to determine the influence of fluctuations of parameters such as the mutation strength on the performance. Such an analysis is necessary for determining the influence of the cumulation parameter c and of the damping constant D on the performance of the algorithm and thus for a complete understanding of cumulative mutation strength adaptation on the noisy sphere.

In Chapter 6, we have empirically compared the efficiency of the $(\mu/\mu, \lambda)$-ES with cumulative mutation strength adaptation with that of a number of other search strategies. As a testing environment, we used noisy spheres with different search space dimensionalities and fitness-proportionate noise strengths. We have found that except for a gradient-based strategy in the absence of noise, all strategies we considered at best exhibit what could be called "noisy linear convergence". Due to the scale invariance of the environment, the graph of the logarithm of the function values over the number of objective function evaluations looks like a straight line with some superimposed fluctuations. The negative of the slope of that line determines the efficiency of a strategy. We observed that with the exceptions of the simplex method that is highly unstable and the multi-directional search method that does not scale well with the search space dimensionality, for all of the strategies we considered the efficiencies are of the same order of magnitude. Whether new candidate solutions were generated deterministically or stochastically, or how many candidate solutions were generated per time step had a merely quantitative influence on the efficiencies at least in the absence of noise. However, qualitative differences between the strategies were observed with regard to their robustness. While above a certain noise strength almost all strategies failed to achieve noisy linear convergence, where failure was usually marked by a rapid decrease in step lengths and correspondingly by virtual stagnation of the search, the point above which this happened differed widely for the various strategies. Especially for high search space dimensionalities, the $(\mu/\mu, \lambda)$-ES with cumulative mutation strength adaptation fared best of all of the strategies we tried. At the same time, the cumulative mutation strength adaptation scheme is the only method considered here that explicitly accumulates information for the adaptation of step lengths over a number of time steps. The averaging that is inherent in the accumulation of the progress vectors seems to make cumulative mutation strength adaptation relatively robust with regard to the effects of noise. Therefore, we believe that it is of interest to attempt to adapt the cumulative mutation strength adaptation mechanism to other strategies such as the evolutionary gradient search method. Also of interest is a study of the robustness in the presence of noise of other mutation strength adaptation schemes for evolutionary algorithms, such as mutative self-adaptation.

Finally, it is important to keep in mind that all results have been obtained in linear or spherically symmetric fitness environments. The work presented here can be extended both to other fitness environments and to other types of strategies. A straightforward continuation of the path taken in this book would be to study the performance of evolution strategies on a noisy version of the ridge that was studied in the absence of noise by Oyman et al. [61] in an attempt to model features of fitness landscapes in greater distance from the optimizer. Such an analysis does not require any tools beyond those employed in the present work. Analyses of the performance of evolution strategies in fitness environments that do not exhibit perfect symmetries on the other hand will require new tools such as the differential geometric approach suggested by Beyer [19]. Strategy variants that have been left unconsidered so far and that are of interest include the $(\mu/2, \lambda)$-ES — possibly with mutative self-adaptation — and nested evolution strategies. The analysis of their behavior in fitness environments such as the noisy sphere would be a first step towards a quantitative understanding of their scaling properties.

Appendices

Appendix A
Some Statistical Basics

In this appendix, we list some important concepts, definitions, and elementary results from probability theory and statistics that are used throughout the present work. In particular, we discuss moments and cumulants as important characteristics of probability distributions, introduce expansions of probability distributions, and outline some very basic concepts from the area of order statistics. Clearly, the issues pertaining to those topics far exceed what can be presented in just a few pages. This appendix is therefore rather sketchy. For a much more careful and complete account of the field, we refer to Stuart and Ord [84]. A good introduction to probability theory is provided by Rényi [68]. Many of the important concepts are also presented very readably by Walpole and Myers [89]. Our goal in including this appendix in this book is twofold. First, we feel it is useful for establishing a consistent notation that is used throughout this work. And second, we would like to see it used as a first reference, much in the spirit of Weisstein [90], that can help the reader with some of the concepts he or she is not immediately familiar with, and make it possible to read through this book without the need to consult further literature.

1. Random Variables and Probability Distributions

A *statistical experiment* is any process that generates a set of data. The set of all possible outcomes of a statistical experiment is called the *sample space*. A *random variable* is a function that associates a real number with each element in the sample space. The value assumed by the random variable is thus determined by the outcome of the experiment. The process of sampling the random variable, i.e. of conducting a statistical experiment and observing the value of the random variable, is called a *trial*.

A random variable X is to be regarded as defined by a *distribution function* $P(x)$. The distribution function, sometimes also called the *probability distribution function* or *cumulative distribution function*, gives the probability that a trial yields a value of the random variable less than or equal to x. Thus, the probability distribution function is a non-decreasing function with $P(-\infty) = 0$ and $P(\infty) = 1$. Throughout this appendix as well as in most of this entire work we are concerned with continuous random variables only, i.e. with random variables with continuous distribution functions. Unless noted otherwise, "random variable" or "variate" therefore always refer to continuous random variables.

The distribution function of a random variable is related to its *probability density function* $p(x)$ by

$$P(x) = \int_{-\infty}^{x} p(u)\mathrm{d}u.$$

Equivalently, the probability density function of a random variable equals the derivative of its distribution function. The product $p(x)\mathrm{d}x$ indicates the probability that the value of the random variable is in an interval of width $\mathrm{d}x$ centered at x. Clearly, the properties

$$p(x) \geq 0 \text{ for all } x \in I\!R \qquad \text{and} \qquad \int_{-\infty}^{\infty} p(x)\mathrm{d}x = 1.$$

hold for any probability density function.

Now consider two random variables X and Y. The *joint distribution function* $P(x, y)$ of X and Y gives the joint probability that a trial yields a value of X that is less than or equal to x and a value of Y that is less than or equal to y. It is a non-decreasing function of each of its arguments with $P(x, -\infty) = P(-\infty, y) = 0$ and $P(\infty, \infty) = 1$. The joint distribution function is related to the *joint probability density function* $p(x, y)$ by

$$P(x, y) = \int_{-\infty}^{x} \int_{-\infty}^{y} p(u, v)\mathrm{d}v\mathrm{d}u.$$

The product $p(x, y)\mathrm{d}x\mathrm{d}y$ thus indicates the probability that the values of X and Y are in intervals of widths $\mathrm{d}x$ and $\mathrm{d}y$ centered at x and y, respectively. Clearly, the properties

$$p(x, y) \geq 0 \text{ for all } x, y \in I\!R \qquad \text{and} \qquad \int_{-\infty}^{\infty} \int_{-\infty}^{\infty} p(x, y)\mathrm{d}y\mathrm{d}x = 1$$

hold for any joint probability density function. The random variables X and Y are said to be *independent* if their joint probability density function equals the product of the individual probability density functions. The *marginal distributions* of X and Y are the univariate distributions with distribution functions $P(x, \infty)$ and $P(\infty, y)$. The corresponding probability density functions are

$$p_X(x) = \int_{-\infty}^{\infty} p(x, y)\mathrm{d}y \qquad \text{and} \qquad p_Y(y) = \int_{-\infty}^{\infty} p(x, y)\mathrm{d}x.$$

The generalization of the concepts of joint distribution functions, joint probability density functions, independence, and marginal distributions to more than two random variables is straightforward.

Finally, we will be concerned with sums of random variables on several occasions. The sum of two independent random variables X and Y with density functions $p_1(x)$ and $p_2(y)$ is itself a random variable with density function

$$\begin{aligned} p(z) &= \int_{-\infty}^{\infty} p_1(x)p_2(z-x)\mathrm{d}x \\ &= \int_{-\infty}^{\infty} p_1(z-y)p_2(y)\mathrm{d}y. \end{aligned}$$

We refer to Rényi [68] for the easy proof.

2. Moments and Cumulants

Moments and cumulants are important characteristics of probability distributions that play a prominent role in probability theory and in statistics. In this section, we briefly outline the most important definitions and interdependencies.

2.1 Moments

Moments are descriptive constants that can be useful for measuring properties of and, in certain circumstances, for specifying probability distributions. The kth *moment about zero* of a probability distribution $P(x)$ with corresponding density function $p(x)$, where k is a positive integer, is

$$\mu'_k = \int_{-\infty}^{\infty} x^k p(x)\mathrm{d}x. \tag{A.1}$$

The first moment about zero μ'_1 is referred to as the *mean* or the *expected value* of the distribution. As we can identify random variables with their probability distributions, we also talk of the mean of a random variable X and write $\mathrm{E}[X]$ for μ'_1 .

Moments about the mean, or *central moments*, of the distribution are defined for $k \geq 2$ as

$$\mu_k = \int_{-\infty}^{\infty} \left(x - \mu'_1\right)^k p(x)\mathrm{d}x.$$

The second central moment μ_2 is referred to as the *variance* of the distribution. We occasionally write $\mathrm{Var}[X]$ for the variance of random variable X and refer to its square root as the *standard deviation*. The third and fourth central moments are related to the *skewness*

$$\gamma_1 = \frac{\mu_3}{\mu_2^{3/2}} \tag{A.2}$$

and the *kurtosis*

$$\gamma_2 = \frac{\mu_4}{\mu_2^2} - 3 \tag{A.3}$$

of the distribution that quantify the degrees of asymmetry and peakedness, respectively. The central moments can be expressed in terms of the moments about zero as

$$\left.\begin{aligned} \mu_2 &= \mu'_2 - {\mu'_1}^2 \\ \mu_3 &= \mu'_3 - 3\mu'_2\mu'_1 + 2{\mu'_1}^3 \\ \mu_4 &= \mu'_4 - 4\mu'_3\mu'_1 + 6\mu'_2{\mu'_1}^2 - 3{\mu'_1}^4 \\ &\vdots \end{aligned}\right\} \tag{A.4}$$

An alternative to using Equation (A.1) for computing the moments about zero of a probability distribution is offered by the *characteristic function*. The characteristic function of a probability distribution is the Fourier transform

$$\chi(t) = \int_{-\infty}^{\infty} \mathrm{e}^{itx} p(x)\mathrm{d}x$$

of its probability density function, where i denotes the imaginary unit with $i^2 = -1$. Under certain general conditions, the characteristic function determines and is completely determined by the distribution function. The Taylor expansion around zero of the characteristic function is

$$\chi(t) = \sum_{k=0}^{\infty} \frac{(it)^k}{k!}\mu'_k,$$

where $\mu'_0 = 1$. That is, if $\chi(t)$ is expanded in powers of t, then μ'_k must be equal to the coefficient of $(it)^k/k!$ in the expansion. It follows that

$$\mu'_k = (-i)^k \left.\frac{\mathrm{d}^k}{\mathrm{d}t^k}\chi(t)\right|_{t=0}.$$

Thus, if the characteristic function of a distribution is known, then its moments can be determined simply by computing its derivatives at $t = 0$.

2.2 Cumulants

Cumulants are an alternative set of characteristic constants of a probability distribution that in many occasions have more favorable properties than the moments. The cumulants are defined via the *cumulant generating function*

$$\psi(t) = \log(\chi(t)),$$

which is simply the natural logarithm of the characteristic function. Expanding the cumulant generating function yields

$$\psi(t) = \sum_{k=1}^{\infty} \frac{(it)^k}{k!} \kappa_k .$$

That is, if $\psi(t)$ is expanded in powers of t, then κ_k is equal to the coefficient of $(it)^k/k!$ in the expansion. It follows that

$$\kappa_k = (-i)^k \left. \frac{\mathrm{d}^k}{\mathrm{d}t^k} \psi(t) \right|_{t=0} .$$

Alternatively, the cumulants can be expressed in terms of the moments as

$$\left. \begin{aligned} \kappa_1 &= \mu_1' \\ \kappa_2 &= \mu_2 \\ \kappa_3 &= \mu_3 \\ \kappa_4 &= \mu_4 - 3\mu_2^2 \\ &\vdots \end{aligned} \right\} \qquad \text{(A.5)}$$

Note that the first cumulant is the mean of the distribution, and that all cumulants but the first depend on central moments of the distribution only. Cumulants possess transformation properties that make them an attractive choice for many purposes. All cumulants but the first are invariant under a change of origin. If the variate values are multiplied by a constant c, then the kth cumulant κ_k is multiplied by c^k.

2.3 Sums of Independent Random Variables

The characteristic function of a sum of independent random variables is the product of the individual characteristic functions. Thus, the cumulant generating function of a sum of independent random variables equals the sum of the individual cumulant generating functions. It follows that the kth cumulant of a sum of independent random variables is the sum of the kth cumulants of its summands. Therefore, denoting the kth central moments of independent random variables X and Y by $\mu_k^{(X)}$ and $\mu_k^{(Y)}$, respectively, it follows from Equations (A.5) that the central moments of the sum $X + Y$ are

$$\left. \begin{aligned} \mu_2(X+Y) &= \mu_2(X) + \mu_2(Y) \\ \mu_3(X+Y) &= \mu_3(X) + \mu_3(Y) \\ \mu_4(X+Y) &= \mu_4(X) + \mu_4(Y) + 6\mu_2(X)\mu_2(Y) \\ &\vdots \end{aligned} \right\} \qquad \text{(A.6)}$$

Note that the central moments of the sum of independent random variables are functions of the central moments of the summands only.

3. Some Important Distributions

Two probability distributions that are of particular importance throughout this work are the normal distribution and the χ^2-distribution.

3.1 The Normal Distribution

The *normal distribution* is arguably the most important distribution in statistics. It derives much of its significance from the *central limit theorem* that states that the mean of any set of variates independently drawn from any distribution having finite mean and variance tends to a normal distribution. A random variable is called *normally distributed* if its probability density function has the form

$$p(x) = \frac{1}{\sqrt{2\pi}\sigma} \exp\left(-\frac{1}{2}\left(\frac{x-\mu}{\sigma}\right)^2\right), \qquad -\infty < x < \infty.$$

The parameters μ and σ are the mean and the standard deviation of the distribution, respectively. The corresponding distribution function is

$$P(x) = \frac{1}{2}\left(1 + \operatorname{erf}\left(\frac{x-\mu}{\sqrt{2}\sigma}\right)\right),$$

where erf denotes the error function. The characteristic function of the normal distribution is

$$\chi(t) = \exp\left(i\mu t - \frac{1}{2}\sigma^2 t^2\right).$$

It can easily be seen that the central moments of the normal distribution are

$$\mu_k = \begin{cases} \dfrac{k!}{2^{k/2}(k/2)!}\sigma^k & \text{if } k \text{ is even} \\ 0 & \text{if } k \text{ is odd} \end{cases}, \qquad k \geq 2.$$

The cumulants of the normal distribution are $\kappa_1 = \mu$, $\kappa_2 = \sigma^2$, and $\kappa_k = 0$ for $k > 2$.

Of particular importance is the standardized normal distribution with mean zero and with variance one. Throughout this work, we use the symbols Φ and ϕ to refer to the distribution function and the probability density function of the standardized normal distribution, respectively.

3.2 The χ^2-Distribution

A random variable is called χ_n^2*-distributed* if its probability density function is of the form

$$p(x) = \frac{1}{2^{n/2}\Gamma(n/2)} x^{n/2-1} e^{-\frac{1}{2}x^2}, \qquad 0 \leq x < \infty.$$

The sum of squares of n independent standardized normal variates is χ_n^2-distributed. The parameter n is referred to as the number of degrees of freedom of the distribution. The characteristic function of the χ_n^2-distribution is

$$\chi(t) = (1 - 2it)^{-\frac{1}{2}n}.$$

It can be seen that the mean of the distribution is $\mu_1' = n$ and that the central moments are $\mu_2 = 2n$, $\mu_3 = 8n$, $\mu_4 = 12n(n+4)$, ... The kth cumulant is $\kappa_k = 2^{k-1}(k-1)!n$. As a consequence of the central limit theorem, as n tends to infinity the χ_n^2-distribution tends to normality.

4. Expansions of Probability Distributions

Gram-Charlier expansions and Cornish-Fisher expansions represent probability density functions and inverse distribution functions in terms of normal distributions.

4.1 Hermite Polynomials

Hermite polynomials occur naturally in connection with successive derivatives of the probability density function of the standardized normal distribution and are defined by the identity

$$\frac{\mathrm{d}^k}{\mathrm{d}x^k}\phi(x) = (-1)^k \mathrm{He}_k(x)\phi(x),$$

where $\phi(x)$ is the probability density function of the standardized normal distribution defined in Section 3.1. Explicit calculation shows that the first Hermite polynomials are

$$\begin{aligned}
\mathrm{He}_0(x) &= 1\\
\mathrm{He}_1(x) &= x\\
\mathrm{He}_2(x) &= x^2 - 1\\
\mathrm{He}_3(x) &= x^3 - 3x\\
\mathrm{He}_4(x) &= x^4 - 6x^2 + 3\\
\mathrm{He}_5(x) &= x^5 - 10x^3 + 15x\\
\mathrm{He}_6(x) &= x^6 - 15x^4 + 45x^2 - 15\\
\mathrm{He}_7(x) &= x^7 - 21x^5 + 105x^3 - 105x\\
\mathrm{He}_8(x) &= x^8 - 28x^6 + 210x^4 - 420x^2 + 105\\
&\vdots
\end{aligned}$$

Generally, the kth Hermite polynomial $\mathrm{He}_k(x)$ is of degree k in x and the coefficient of x^k is unity. Furthermore, it can be shown that for $k \geq 1$ the identities

$$\frac{\mathrm{d}}{\mathrm{d}x}\mathrm{He}_k(x) = k\mathrm{He}_{k-1}(x) \tag{A.7}$$

and

$$\mathrm{He}_{k+1}(x) = x\mathrm{He}_k(x) - k\mathrm{He}_{k-1}(x) \tag{A.8}$$

hold. Moreover, the orthogonality property

$$\int_{-\infty}^{\infty} \mathrm{He}_k(x)\mathrm{He}_l(x)\phi(x)\mathrm{d}x = \delta_{kl}k!, \tag{A.9}$$

where δ_{kl} denotes the Kronecker delta, holds for any $k, l \geq 0$.

4.2 Gram-Charlier Expansions

Gram-Charlier expansions seek to represent a density function $p(x)$ as a series in the derivatives of a normal density function. The approach relies on the expectation that if two distributions have a certain number of cumulants in common, they will bear some resemblance to each other. If cumulants up to order k are identical, it is often the case that as k tends to infinity the distributions approach each other. Thus, the hope is that by equating the lower-order moments of two distributions they can be brought to approximate equality.

Let us assume for the moment that the distribution the density of which is to be approximated is standardized. The expansion is then in terms of the standardized normal density and reads

$$p(x) = \sum_{k=0}^{\infty} (-1)^k c_k \frac{\mathrm{d}^k}{\mathrm{d}x^k} \phi(x). \tag{A.10}$$

The coefficients c_k can be identified by means of multiplying the series with $\mathrm{He}_k(x)$, integrating from $-\infty$ to ∞, and using Equation (A.9). Inserting the results of the calculations in Equation (A.10), the beginning of the series reads

$$p(x) = \left[1 + \frac{\kappa_3}{3!}\mathrm{He}_3(x) + \frac{\kappa_4}{4!}\mathrm{He}_4(x) + \frac{\kappa_5}{5!}\mathrm{He}_5(x) + \frac{\kappa_6 + 10\kappa_3^2}{6!}\mathrm{He}_6(x) + \dots\right]\phi(x), \tag{A.11}$$

where the κ_k are the cumulants of the distribution the density of which is to be approximated. Uniform convergence of the series can be shown under certain general conditions. A rearrangement of the terms of the Gram-Charlier series that has better asymptotic convergence properties provided certain scaling properties of the cumulants has come to be known as *Edgeworth expansion*. We refer to Stuart and Ord [84] for details.

If the distribution the density function of which is to be approximated is not standardized, the expansion is in terms of derivatives of the normal distribution with the mean and variance of the distribution to be expanded and reads

$$p(x) = \frac{1}{\sqrt{\kappa_2}}\left[1 + \frac{\gamma_1}{3!}\mathrm{He}_3\left(\frac{x-\kappa_1}{\sqrt{\kappa_2}}\right) + \frac{\gamma_2}{4!}\mathrm{He}_4\left(\frac{x-\kappa_1}{\sqrt{\kappa_2}}\right) + \dots\right]\phi\left(\frac{x-\kappa_1}{\sqrt{\kappa_2}}\right), \tag{A.12}$$

where again the κ_k are the cumulants of the distribution the density of which is to be expanded, and where γ_1 and γ_2 are the skewness and kurtosis defined in Equations (A.2) and (A.3), respectively.

In practical applications, only a finite number of terms in the expansion can be considered. The important question is not whether an infinite series can represent a density function, but whether a finite number of terms can serve as a satisfactory approximation. While this is often the case, it has to be kept in mind that the sum of a finite number of terms may give negative values, and that the sum may behave irregularly in the sense that the sum of k terms may give a worse fit than the sum of $k-1$ terms. The finite series approach is useful only in cases of moderate skewness of the distribution, and it is of little use if it is the tails of the distribution that are of interest.

4.3 Cornish-Fisher Expansions

The *Cornish-Fisher expansion* is an expansion of the inverse function of an arbitrary distribution function. Let $P(y)$ be a standardized distribution function with cumulants κ_k. Suppose that x is defined by

$$P(y) = \Phi(x).$$

Then, y can be expressed as a polynomial in x by expanding both sides of the equation in terms of derivatives of the standardized normal distribution and by subsequently rearranging terms. The result of the tedious calculations can be found for example in Abramowitz and Stegun [1] and in Fisher and Cornish [36] and reads

$$\begin{aligned} y &= P^{-1}(\Phi(x)) \\ &= x + \frac{\kappa_3}{3!}\mathrm{He}_2(x) + \frac{\kappa_4}{4!}\mathrm{He}_3(x) - \frac{\kappa_3^2}{(3!)^2}(2\mathrm{He}_3(x) + \mathrm{He}_1(x)) + \dots \end{aligned} \tag{A.13}$$

If the distribution function the inverse of which is to be expanded is not standardized, the result reads

$$P^{-1}(\Phi(x)) = \kappa_1 + \kappa_2^{1/2}\left[x + \frac{\gamma_1}{3!}\mathrm{He}_2(x) + \frac{\gamma_2}{4!}\mathrm{He}_3(x) + \dots\right], \tag{A.14}$$

where γ_1 and γ_2 are the skewness and the kurtosis of the distribution, respectively.

5. Order Statistics

Given a sample of n independent values $x_1, \dots, x_n$ from a probability distribution $P(x)$, the variate with the kth smallest value is referred to as the kth order statistic of the sample. Its distribution function is

$$P_{k:n}(x) = I_{P(x)}(k, n-k+1),$$

where

$$\begin{aligned} I_y(a,b) &= \frac{(a+b-1)!}{(a-1)!(b-1)!}\int_0^y x^{a-1}(1-x)^{b-1}\,\mathrm{d}x \\ &= \sum_{i=1}^{b}\binom{a+b-1}{i-1}x^{a+b-i}(1-x)^{i-1} \end{aligned} \tag{A.15}$$

is the incomplete regularized Beta function, and where the different representations have been derived from Equations **6.6.4** and **26.5.1** in Abramowitz and Stegun [1]. By means of the substitution $x = P(y)$, the probability density function of the kth order statistic can be seen to be

$$p_{k:n}(x) = \frac{n!}{(n-k)!(k-1)!}p(x)[P(x)]^{k-1}[1-P(x)]^{n-k}. \tag{A.16}$$

Likewise, the joint probability density function of the kth and the lth order statistics with $1 \leq k < l \leq n$ is

$$\begin{aligned} p_{k,l:n}(x,y) = &\frac{n!}{(n-l)!(l-k-1)!(k-1)!}p(x)p(y) \\ &[P(x)]^{k-1}[P(y)-P(x)]^{l-k-1}[1-P(y)]^{n-l}. \end{aligned} \tag{A.17}$$

The extension to joint probability density functions of more than two order statistics is straightforward. Much more information on the topic has been assembled by Arnold, Balakrishnan, and Nagaraja [5].

Appendix B
Some Useful Identities

Identity 1

For integer $\nu > 0$ and for any integers μ and λ with $\nu \leq \mu < \lambda$ and real numbers Q_k, $k = 1, \ldots, \nu$, the identity

$$\sum_{i_1=\nu}^{\mu} \sum_{i_2=\nu-1}^{i_1-1} \cdots \sum_{i_\nu=1}^{i_{\nu-1}-1} \prod_{k=0}^{\nu} \frac{[Q_{k+1} - Q_k]^{i_k - i_{k+1} - 1}}{(i_k - i_{k+1} - 1)!}$$
$$= \frac{1}{(\lambda - \mu - 1)!(\mu - \nu)!} \int_0^{Q_1} z^{\lambda-\mu-1}[1 - z]^{\mu-\nu} \mathrm{d}z,$$

where formally $Q_0 = 0$, $Q_{\nu+1} = 1$, $i_0 = \lambda + 1$, and $i_{\nu+1} = 0$ are assumed, holds.

Proof: By induction. Let us write $lhs_{\mu,\lambda}(\nu)$ and $rhs_{\mu,\lambda}(\nu)$ for the left and right hand sides of the identity, respectively. Then,

$$lhs_{\mu,\lambda}(1) = \sum_{i=1}^{\mu} \frac{Q^{\lambda-i}[1 - Q]^{i-1}}{(\lambda - i)!(i - 1)!}.$$

According to Equation (A.15) it follows in terms of the incomplete regularized Beta function that

$$\begin{aligned} lhs_{\mu,\lambda}(1) &= \frac{1}{(\lambda - 1)!} I_Q(\lambda - \mu, \mu) \\ &= \frac{1}{(\lambda - \mu - 1)!(\mu - 1)!} \int_0^{Q} z^{\lambda-\mu-1}[1 - z]^{\mu-1} \mathrm{d}z \\ &= rhs_{\mu,\lambda}(1), \end{aligned}$$

and the validity of the identity for $\nu = 1$ is shown for any μ and λ with $1 \leq \mu < \lambda$.

For the inductive step, let us now assume that the identity holds for a particular value of ν and all values of μ and λ such that $\nu \leq \mu < \lambda$. We need to show that the identity holds for $\nu + 1$.

Let μ and λ satisfy $\nu + 1 \leq \mu < \lambda$. The left hand side of the identity for $\nu + 1$ reads

$$lhs_{\mu,\lambda}(\nu+1) = \sum_{i_1=\nu+1}^{\mu} \sum_{i_2=\nu}^{i_1-1} \cdots \sum_{i_{\nu+1}=1}^{i_\nu-1} \prod_{k=0}^{\nu+1} \frac{[Q_{k+1} - Q_k]^{i_k - i_{k+1} - 1}}{(i_k - i_{k+1} - 1)!},$$

where $Q_0 = 0$, $Q_{\nu+2} = 1$, $i_0 = \lambda + 1$, and $i_{\nu+2} = 0$. Introducing new indices $j_k = i_k - 1$ it follows that

$$lhs_{\mu,\lambda}(\nu+1) = \sum_{j_1=\nu}^{\mu-1} \sum_{j_2=\nu-1}^{j_1-1} \cdots \sum_{j_{\nu+1}=0}^{j_\nu-1} \prod_{k=0}^{\nu+1} \frac{[Q_{k+1} - Q_k]^{j_k - j_{k+1} - 1}}{(j_k - j_{k+1} - 1)!},$$

where $j_0 = \lambda$ and $j_{\nu+2} = -1$. The innermost sum is

$$\begin{aligned}
&\sum_{j_{\nu+1}=0}^{j_\nu-1} \prod_{k=0}^{\nu+1} \frac{[Q_{k+1} - Q_k]^{j_k - j_{k+1} - 1}}{(j_k - j_{k+1} - 1)!} \\
&= \left[\prod_{k=0}^{\nu-1} \frac{[Q_{k+1} - Q_k]^{j_k - j_{k+1} - 1}}{(j_k - j_{k+1} - 1)!}\right] \sum_{j_{\nu+1}=0}^{j_\nu-1} \frac{[Q_{\nu+1} - Q_\nu]^{j_\nu - j_{\nu+1} - 1}[1 - Q_{\nu+1}]^{j_{\nu+1}}}{(j_\nu - j_{\nu+1} - 1)! j_{\nu+1}!} \\
&= \left[\prod_{k=0}^{\nu-1} \frac{[Q_{k+1} - Q_k]^{j_k - j_{k+1} - 1}}{(j_k - j_{k+1} - 1)!}\right] \frac{[1 - Q_\nu]^{j_\nu - 1}}{(j_\nu - 1)!} \\
&= \prod_{k=0}^{\nu} \frac{[Q_{k+1} - Q_k]^{j_k - j_{k+1} - 1}}{(j_k - j_{k+1} - 1)!},
\end{aligned}$$

where the second step follows from the binomial theorem and where in the last line $Q_{\nu+1} = 1$ and $j_{\nu+1} = 0$. It thus follows that

$$\begin{aligned}
lhs_{\mu,\lambda}(\nu+1) &= \sum_{j_1=\nu}^{\mu-1} \sum_{j_2=\nu-1}^{j_1-1} \cdots \sum_{j_\nu=1}^{j_{\nu-1}-1} \prod_{k=0}^{\nu} \frac{[Q_{k+1} - Q_k]^{j_k - j_{k+1} - 1}}{(j_k - j_{k+1} - 1)!} \\
&= lhs_{\mu-1,\lambda-1}(\nu).
\end{aligned}$$

As $\nu \leq \mu - 1 < \lambda - 1$ and as the identity holds for ν it follows that

$$\begin{aligned}
lhs_{\mu,\lambda}(\nu+1) &= rhs_{\mu-1,\lambda-1}(\nu) \\
&= \frac{1}{(\lambda - \mu - 1)!(\mu - \nu - 1)!} \int_0^{Q_1} z^{\lambda-\mu-1}[1 - z]^{\mu-\nu-1} \mathrm{d}z \\
&= rhs_{\mu,\lambda}(\nu+1)
\end{aligned}$$

and the validity of the identity has been shown. □

Identity 2

For non-negative integer k and real numbers E and D, the identity

$$\frac{1}{\sqrt{2\pi}}\int_{-\infty}^{\infty}\mathrm{He}_k(x)\mathrm{e}^{-\frac{1}{2}x^2}\exp\left(-\frac{1}{2}\left(\frac{x-E}{D}\right)^2\right)\mathrm{d}x$$
$$=\frac{D}{(1+D^2)^{(k+1)/2}}\mathrm{He}_k\left(\frac{E}{(1+D^2)^{1/2}}\right)\exp\left(-\frac{1}{2}\frac{E^2}{1+D^2}\right)$$

holds.

Proof: By induction. It is easily verified that by means of quadratic completion,

$$x^2+\left(\frac{x-E}{D}\right)^2=\frac{x^2(1+D^2)-2Ex+E^2}{D^2}$$
$$=\left(\frac{x-E/(1+D^2)}{D/(1+D^2)^{1/2}}\right)^2+\frac{E^2}{1+D^2}. \tag{B.1}$$

Let us write lhs_k and rhs_k for the left and right hand sides of the identity, respectively. Then, for $k=0$ and using Equation (B.1) and subsequently the substitution $y=(x-E/(1+D^2))/(D/(1+D^2)^{1/2})$,

$$\begin{aligned}
lhs_0 &= \frac{1}{\sqrt{2\pi}}\int_{-\infty}^{\infty}\mathrm{e}^{-\frac{1}{2}x^2}\exp\left(-\frac{1}{2}\left(\frac{x-E}{D}\right)^2\right)\mathrm{d}x\\
&= \frac{1}{\sqrt{2\pi}}\exp\left(-\frac{1}{2}\frac{E^2}{1+D^2}\right)\int_{-\infty}^{\infty}\exp\left(-\frac{1}{2}\left(\frac{x-E/(1+D^2)}{D/(1+D^2)^{1/2}}\right)^2\right)\mathrm{d}x\\
&= \frac{1}{\sqrt{2\pi}}\frac{D}{(1+D^2)^{1/2}}\exp\left(-\frac{1}{2}\frac{E^2}{1+D^2}\right)\int_{-\infty}^{\infty}\mathrm{e}^{-\frac{1}{2}y^2}\mathrm{d}y\\
&= \frac{D}{(1+D^2)^{1/2}}\exp\left(-\frac{1}{2}\frac{E^2}{1+D^2}\right)\\
&= rhs_0.
\end{aligned}$$

Similarly, for $k=1$,

$$\begin{aligned}
lhs_1 &= \frac{1}{\sqrt{2\pi}}\int_{-\infty}^{\infty}x\mathrm{e}^{-\frac{1}{2}x^2}\exp\left(-\frac{1}{2}\left(\frac{x-E}{D}\right)^2\right)\mathrm{d}x\\
&= \frac{1}{\sqrt{2\pi}}\exp\left(-\frac{1}{2}\frac{E^2}{1+D^2}\right)\int_{-\infty}^{\infty}x\exp\left(-\frac{1}{2}\left(\frac{x-E/(1+D^2)}{D/(1+D^2)^{1/2}}\right)^2\right)\mathrm{d}x\\
&= \frac{1}{\sqrt{2\pi}}\frac{D}{(1+D^2)^{1/2}}\exp\left(-\frac{1}{2}\frac{E^2}{1+D^2}\right)\int_{-\infty}^{\infty}\left(\frac{E}{1+D^2}+\frac{Dy}{(1+D^2)^{1/2}}\right)\mathrm{e}^{-\frac{1}{2}y^2}\mathrm{d}y\\
&= \frac{D}{1+D^2}\frac{E}{(1+D^2)^{1/2}}\exp\left(-\frac{1}{2}\frac{E^2}{1+D^2}\right)\\
&= rhs_1.
\end{aligned}$$

Given its validity for $k-1$ and k, where $k \geq 1$, the validity of the identity for $k+1$ can be shown by computing the derivatives of lhs_k and rhs_k with respect to E. Using Equation (A.8),

$$\begin{aligned}
\frac{\mathrm{d}}{\mathrm{d}E} lhs_k &= \frac{1}{\sqrt{2\pi}} \int_{-\infty}^{\infty} \mathrm{He}_k(x) \mathrm{e}^{-\frac{1}{2}x^2} \frac{\mathrm{d}}{\mathrm{d}E} \exp\left(-\frac{1}{2}\left(\frac{x-E}{D}\right)^2\right) \mathrm{d}x \\
&= \frac{1}{\sqrt{2\pi}} \int_{-\infty}^{\infty} \frac{x-E}{D^2} \mathrm{He}_k(x) \mathrm{e}^{-\frac{1}{2}x^2} \exp\left(-\frac{1}{2}\left(\frac{x-E}{D}\right)^2\right) \mathrm{d}x \\
&= \frac{1}{\sqrt{2\pi} D^2} \int_{-\infty}^{\infty} [\mathrm{He}_{k+1}(x) + k\mathrm{He}_{k-1}(x) - E\mathrm{He}_k(x)] \\
&\qquad\qquad \mathrm{e}^{-\frac{1}{2}x^2} \exp\left(-\frac{1}{2}\left(\frac{x-E}{D}\right)^2\right) \mathrm{d}x \\
&= \frac{1}{D^2} lhs_{k+1} + \frac{k}{D^2} lhs_{k-1} - \frac{E}{D^2} lhs_k.
\end{aligned} \tag{B.2}$$

Similarly, using Equation (A.7),

$$\begin{aligned}
\frac{\mathrm{d}}{\mathrm{d}E} rhs_k &= \frac{D}{(1+D^2)^{(k+1)/2}} \frac{\mathrm{d}}{\mathrm{d}E} \left[\mathrm{He}_k\left(\frac{E}{(1+D^2)^{1/2}}\right) \exp\left(-\frac{1}{2}\frac{E^2}{1+D^2}\right)\right] \\
&= \frac{D}{(1+D^2)^{(k+1)/2}} \left[\frac{k}{(1+D^2)^{1/2}} \mathrm{He}_{k-1}\left(\frac{E}{(1+D^2)^{1/2}}\right)\right. \\
&\qquad \left. - \frac{E}{1+D^2} \mathrm{He}_k\left(\frac{E}{(1+D^2)^{1/2}}\right)\right] \exp\left(-\frac{1}{2}\frac{E^2}{1+D^2}\right) \\
&= \frac{k}{1+D^2} rhs_{k-1} - \frac{E}{1+D^2} rhs_k.
\end{aligned} \tag{B.3}$$

As the identity holds for k and for all values of E, it follows that $\mathrm{d}lhs_k/\mathrm{d}E = \mathrm{d}rhs_k/\mathrm{d}E$. Therefore, from Equations (B.2) and (B.3) it follows that

$$lhs_{k+1} = Elhs_k - klhs_{k-1} + \frac{D^2 k}{1+D^2} rhs_{k-1} - \frac{D^2 E}{1+D^2} rhs_k.$$

Using the identity for k and for $k-1$, it follows that

$$lhs_{k+1} = \frac{E}{1+D^2} rhs_k - \frac{k}{1+D^2} rhs_{k-1}.$$

The validity of the identity now follows immediately from Equation (A.8). □

Identity 3

For non-negative integer k and real numbers E and D, the identity

$$\frac{1}{\sqrt{2\pi}}\int_{-\infty}^{\infty}\mathrm{He}_k(x)\mathrm{e}^{-\frac{1}{2}x^2}\Phi\left(\frac{x-E}{D}\right)\mathrm{d}x$$
$$=\begin{cases} 1-\Phi\left(\dfrac{E}{(1+D^2)^{1/2}}\right) & \text{if } k=0 \\ \dfrac{1}{\sqrt{2\pi}}\dfrac{1}{(1+D^2)^{k/2}}\mathrm{He}_{k-1}\left(\dfrac{E}{(1+D^2)^{1/2}}\right)\exp\left(-\dfrac{1}{2}\dfrac{E^2}{1+D^2}\right) & \text{if } k>0 \end{cases}$$

holds.

Proof: Follows from integration of Identity 2. Let us write lhs_k and rhs_k for the left and the right hand sides of the current identity, and let l_k and r_k denote the left and right hand sides of Identity 2, respectively. Then, using Equation (B.6),

$$\begin{aligned}\int l_k\mathrm{d}E &= \frac{1}{\sqrt{2\pi}}\int_{-\infty}^{\infty}\mathrm{He}_k(x)\mathrm{e}^{-\frac{1}{2}x^2}\int\exp\left(-\frac{1}{2}\left(\frac{x-E}{D}\right)^2\right)\mathrm{d}E\mathrm{d}x \\ &= D\int_{-\infty}^{\infty}\mathrm{He}_k(x)\mathrm{e}^{-\frac{1}{2}x^2}\left[1-\Phi\left(\frac{x-E}{D}\right)\right]\mathrm{d}x \\ &= \begin{cases}\sqrt{2\pi}D(1-lhs_k) & \text{if } k=0 \\ -\sqrt{2\pi}D\ lhs_k & \text{if } k>0\end{cases}.\end{aligned} \tag{B.4}$$

Similarly, again using Equation (B.6),

$$\begin{aligned}\int r_k\mathrm{d}E &= \frac{D}{(1+D^2)^{(k+1)/2}}\int\mathrm{He}_k\left(\frac{E}{(1+D^2)^{1/2}}\right)\exp\left(-\frac{1}{2}\frac{E^2}{1+D^2}\right)\mathrm{d}E \\ &= \begin{cases}\sqrt{2\pi}D\Phi\left(\dfrac{E}{(1+D^2)^{1/2}}\right) & \text{if } k=0 \\ -\dfrac{D}{(1+D^2)^{k/2}}\mathrm{He}_{k-1}\left(\dfrac{E}{(1+D^2)^{1/2}}\right)\exp\left(-\dfrac{1}{2}\dfrac{E^2}{1+D^2}\right) & \text{if } k>0\end{cases} \\ &= \begin{cases}\sqrt{2\pi}D(1-rhs_k) & \text{if } k=0 \\ -\sqrt{2\pi}D\ rhs_k & \text{if } k>0\end{cases}.\end{aligned} \tag{B.5}$$

As Identity 2 holds for all values of E it follows that $\int l_k\mathrm{d}E=\int r_k\mathrm{d}E$. Therefore, the current identity follows from Equations (B.4) and (B.5). □

Identity 4

For non-negative integer k and real number z, the identity

$$\int_z^\infty \mathrm{He}_k(x)\phi(x)\mathrm{d}x = \begin{cases} 1-\Phi(z) & \text{if } k=0 \\ \mathrm{He}_{k-1}(z)\phi(z) & \text{if } k>0 \end{cases},$$

where ϕ and Φ denote the probability density function and the distribution function of the standardized normal distribution, respectively, holds.

Proof: For $k = 0$, the validity of the identity is obvious. For $k > 1$, it follows from Equations (A.7) and (A.8) that

$$\begin{aligned}\frac{\mathrm{d}}{\mathrm{d}x}\left[-\mathrm{He}_{k-1}(x)\phi(x)\right] &= -[(k-1)\mathrm{He}_{k-2}(x) - x\mathrm{He}_{k-1}(x)]\phi(x) \\ &= \mathrm{He}_k(x)\phi(x),\end{aligned}$$

and therefore that

$$\int \mathrm{He}_k(x)\phi(x)\mathrm{d}x = -\mathrm{He}_{k-1}(x)\phi(x). \tag{B.6}$$

The validity of the identity for $k > 1$ is an immediate consequence of Equation (B.6). □

Identity 5

For positive integer k and real numbers z and $\beta \neq 0$, the identity

$$\int_z^\infty \mathrm{He}_k(x)[\phi(x)]^\beta \mathrm{d}x$$

$$= \begin{cases} \frac{1}{\beta}[\phi(z)]^\beta & \text{if } k = 1 \\ \frac{1}{\beta}\mathrm{He}_{k-1}(z)[\phi(z)]^\beta - \frac{\beta-1}{\beta}(k-1)\int_z^\infty \mathrm{He}_{k-2}(x)[\phi(x)]^\beta \mathrm{d}x & \text{if } k > 1 \end{cases},$$

where ϕ denotes the probability density function of the standardized normal distribution, holds.

Proof: By integration by parts. Let

$$u(x) = [\phi(x)]^{\beta-1} \quad \text{and} \quad v(x) = -\mathrm{He}_{k-1}(x)\phi(x).$$

It follows from Equation (B.6) that

$$u'(x) = -(\beta-1)x[\phi(x)]^{\beta-1} \quad \text{and} \quad v'(x) = \mathrm{He}_k(x)\phi(x).$$

For $k = 1$, integration by parts yields

$$\int_z^\infty x[\phi(x)]^\beta \mathrm{d}x = [\phi(z)]^\beta - (\beta-1)\int_z^\infty x[\phi(x)]^\beta \mathrm{d}x$$

and the identity follows from transferring the integral from the right hand side to the left and subsequently dividing by β. For $k > 1$, integration by parts and using Equation (A.8) yields

$$\begin{aligned}\int_z^\infty \mathrm{He}_k(x)[\phi(x)]^\beta \mathrm{d}x &= \mathrm{He}_{k-1}(z)[\phi(z)]^\beta - (\beta-1)\int_z^\infty x\mathrm{He}_{k-1}(x)[\phi(x)]^\beta \mathrm{d}x \\ &= \mathrm{He}_{k-1}(z)[\phi(z)]^\beta - (\beta-1)\int_z^\infty \mathrm{He}_k(x)[\phi(x)]^\beta \mathrm{d}x \\ &\quad - (\beta-1)(k-1)\int_z^\infty \mathrm{He}_{k-2}(x)[\phi(x)]^\beta \mathrm{d}x.\end{aligned}$$

Transferring the first integral on the right hand side to the left and subsequently dividing by β proves the identity. □

Identity 6

For non-negative integer k and real number z, the identity

$$\int_z^\infty \mathrm{He}_k(x)\phi(x)[1-\Phi(x)]\mathrm{d}x = \begin{cases} \frac{1}{2}[1-\Phi(x)]^2 & \text{if } k=0 \\ \mathrm{He}_{k-1}(z)\phi(z)[1-\Phi(z)] - \int_z^\infty \mathrm{He}_{k-1}(x)[\phi(x)]^2\mathrm{d}x & \text{if } k>0 \end{cases},$$

where ϕ and Φ denote the probability density function and the distribution function of the standardized normal distribution, respectively, holds.

Proof: By integration by parts. For $k = 0$, let

$$u(x) = 1-\Phi(x) \qquad \text{and} \qquad v(x) = \Phi(x).$$

Then,

$$u'(x) = -\phi(x) \qquad \text{and} \qquad v'(x) = \phi(x).$$

Integration by parts yields

$$\begin{aligned} \int_z^\infty \phi(x)[1-\Phi(x)]\mathrm{d}x &= -\Phi(z)[1-\Phi(z)] + \int_z^\infty \phi(x)\Phi(x)\mathrm{d}x \\ &= [1-\Phi(z)]^2 - \int_z^\infty \phi(x)[1-\Phi(x)]\mathrm{d}x. \end{aligned}$$

The identity follows from transferring the integral from the right hand side to the left and subsequently dividing by two.

For $k > 0$, let

$$u(x) = 1-\Phi(x) \qquad \text{and} \qquad v(x) = -\mathrm{He}_{k-1}(x)\phi(x).$$

It follows from Equation (B.6) that

$$u'(x) = -\phi(x) \qquad \text{and} \qquad v'(x) = \mathrm{He}_k(x)\phi(x).$$

Integration by parts immediately yields the identity. □

Identity 7

For positive integer k and real number z, the identity

$$\int_z^\infty \mathrm{He}_k(x)[\phi(x)]^2[1-\Phi(x)]\mathrm{d}x$$

$$= \begin{cases} \frac{1}{2}[\phi(z)]^2[1-\Phi(z)] - \frac{1}{2}\int_z^\infty [\phi(x)]^3\mathrm{d}x & \text{if } k=1 \\ \frac{1}{2}\mathrm{He}_{k-1}(z)[\phi(z)]^2[1-\Phi(z)] - \frac{1}{2}\int_z^\infty \mathrm{He}_{k-1}(x)[\phi(x)]^3\mathrm{d}x & \\ \qquad -\frac{k-1}{2}\int_z^\infty \mathrm{He}_{k-2}(x)[\phi(x)]^2[1-\Phi(x)]\mathrm{d}x & \text{if } k>1 \end{cases},$$

where ϕ and Φ denote the probability density function and the distribution function of the standardized normal distribution, respectively, holds.

Proof: By integration by parts. Let

$$u(x) = \phi(x)[1-\Phi(x)] \qquad \text{and} \qquad v(x) = -\mathrm{He}_{k-1}(x)\phi(x).$$

It follows from Equation (B.6) that

$$u'(x) = -[\phi(x)]^2 - x\phi(x)[1-\Phi(x)] \qquad \text{and} \qquad v'(x) = \mathrm{He}_k(x)\phi(x).$$

Integration by parts yields

$$\begin{aligned}\int_z^\infty \mathrm{He}_k(x)[\phi(x)]^2[1-\Phi(x)]\mathrm{d}x &= \mathrm{He}_{k-1}(z)[\phi(z)]^2[1-\Phi(z)] \\ &\quad - \int_z^\infty \mathrm{He}_{k-1}(x)[\phi(x)]^3\mathrm{d}x \\ &\quad - \int_z^\infty x\mathrm{He}_{k-1}(x)[\phi(x)]^2[1-\Phi(x)]\mathrm{d}x.\end{aligned}$$

For $k=1$, the identity follows by transferring the second integral on the right hand side to the left and subsequently dividing by two. For $k>1$, using Equation (A.8) yields

$$\begin{aligned}\int_z^\infty \mathrm{He}_k(x)[\phi(x)]^2[1-\Phi(x)]\mathrm{d}x &= \mathrm{He}_{k-1}(z)[\phi(z)]^2[1-\Phi(z)] \\ &\quad - \int_z^\infty \mathrm{He}_{k-1}(x)[\phi(x)]^3\mathrm{d}x \\ &\quad - \int_z^\infty \mathrm{He}_k(x)[\phi(x)]^2[1-\Phi(x)]\mathrm{d}x \\ &\quad - (k-1)\int_z^\infty \mathrm{He}_{k-2}(x)[\phi(x)]^2[1-\Phi(x)]\mathrm{d}x.\end{aligned}$$

The identity now follows from transferring the second integral on the right hand side to the left and subsequently dividing by two. □

Appendix C
Computing the Overvaluation

The purpose of this appendix is to provide detailed calculations that have been omitted from Chapter 3 for the sake of readability. The derivations to be presented in what follows occupy some space, but are always straightforward.

1. Preliminaries

Many of the results to be derived in what follows can be expressed conveniently in terms of partial derivatives of the quantity

$$\chi = \Phi\left(\frac{\sigma^{*2}/2+\kappa_1}{\sqrt{\sigma^{*2}+\sigma_\epsilon^{*2}+\kappa_2}}\right)$$

defined in Equation (3.6). It can easily be shown by induction that for $k \geq 1$, the kth derivative of χ with respect to κ_1 is

$$\begin{aligned}\chi^{(k)} &= \frac{\partial^k \chi}{\partial \kappa_1^k} \\ &= \frac{1}{(\sigma^{*2}+\sigma_\epsilon^{*2}+\kappa_2)^{k/2}} \\ &\qquad \mathrm{He}_{k-1}\left(\frac{-\sigma^{*2}/2-\kappa_1}{\sqrt{\sigma^{*2}+\sigma_\epsilon^{*2}+\kappa_2}}\right)\phi\left(\frac{\sigma^{*2}/2+\kappa_1}{\sqrt{\sigma^{*2}+\sigma_\epsilon^{*2}+\kappa_2}}\right).\end{aligned} \tag{C.1}$$

Straightforward calculation shows the validity of Equation (C.1) for $k = 1$. To infer its validity for $k+1$ from its validity for k, it is easily verified that

$$\begin{aligned}\frac{\partial^{k+1}}{\partial \kappa_1^{k+1}}\chi &= \frac{\partial \chi^{(k)}}{\partial \kappa_1} \\ &= \frac{1}{(\sigma^{*2}+\sigma_\epsilon^{*2}+\kappa_2)^{k/2}}\frac{\partial}{\partial \kappa_1}\left[\mathrm{He}_{k-1}\left(\frac{-\sigma^{*2}/2-\kappa_1}{\sqrt{\sigma^{*2}+\sigma_\epsilon^{*2}+\kappa_2}}\right)\right. \\ &\qquad\left.\phi\left(\frac{\sigma^{*2}/2+\kappa_1}{\sqrt{\sigma^{*2}+\sigma_\epsilon^{*2}+\kappa_2}}\right)\right]\end{aligned}$$

$$= \frac{1}{(\sigma^{*2} + \sigma_\epsilon^{*2} + \kappa_2)^{k/2}} \left[\frac{-\sigma^{*2}/2 - \kappa_1}{\sigma^{*2} + \sigma_\epsilon^{*2} + \kappa_2} \mathrm{He}_{k-1} \left(\frac{-\sigma^{*2}/2 - \kappa_1}{\sqrt{\sigma^{*2} + \sigma_\epsilon^{*2} + \kappa_2}} \right) - \frac{k-1}{\sqrt{\sigma^{*2} + \sigma_\epsilon^{*2} + \kappa_2}} \mathrm{He}_{k-2} \left(\frac{-\sigma^{*2}/2 - \kappa_1}{\sqrt{\sigma^{*2} + \sigma_\epsilon^{*2} + \kappa_2}} \right) \right] \phi \left(\frac{\sigma^{*2}/2 + \kappa_1}{\sqrt{\sigma^{*2} + \sigma_\epsilon^{*2} + \kappa_2}} \right)$$

$$= \frac{1}{(\sigma^{*2} + \sigma_\epsilon^{*2} + \kappa_2)^{(k+1)/2}} \mathrm{He}_k \left(\frac{-\sigma^{*2}/2 - \kappa_1}{\sqrt{\sigma^{*2} + \sigma_\epsilon^{*2} + \kappa_2}} \right) \phi \left(\frac{\sigma^{*2}/2 + \kappa_1}{\sqrt{\sigma^{*2} + \sigma_\epsilon^{*2} + \kappa_2}} \right)$$

where Equations (A.7) and (A.8) have been used in the second and third steps, respectively.
As for derivatives with respect to κ_2, straightforward calculation shows that

$$\frac{\partial \chi}{\partial \kappa_2} = \frac{1}{2} \chi^{(2)}.$$

As a simple corollary, for $k \geq 0$,

$$\frac{\partial \chi^{(k)}}{\partial \kappa_2} = \frac{1}{2} \chi^{(k+2)}. \tag{C.2}$$

Furthermore,

$$\frac{\partial \chi}{\partial \sigma^*} = \left(\frac{\partial \kappa_1}{\partial \sigma^*} + \sigma^* \right) \chi^{(1)} + \left(\frac{1}{2} \frac{\partial \kappa_2}{\partial \sigma^*} + \sigma^* \right) \chi^{(2)} \tag{C.3}$$

can be confirmed by straightforward calculation.

2. Obtaining the Distribution

So as to obtain the results quoted in Equations (3.7) and (3.8), we need to first determine probability density $p_{\Xi^*}^{(t+1)}$. Subsequently, mean and variance of the distribution can be obtained easily.

2.1 Computing the Probability Density

The probability density of $\Xi^{*(t+1)}$ is given by Equation (3.3). Using Equations (3.1) and (3.2), for the first integral therein it follows

$$\int_{-\infty}^{\infty} p_{\Xi^*}^{(t)}(y) p_{\xi^*}(x) P_{\xi^*|\Xi^*}^{(acc)}(x|y) \mathrm{d}y$$

$$= \frac{1}{\sigma_\epsilon^* \sqrt{\kappa_2}} \int_{-\infty}^{\infty} \phi \left(\frac{y - \kappa_1}{\sqrt{\kappa_2}} \right) \phi \left(\frac{x}{\sigma_\epsilon^*} \right) \Phi \left(\frac{x - y - \sigma^{*2}/2}{\sigma^*} \right) \mathrm{d}y$$

$$= \frac{1}{\sqrt{2\pi}\sigma_\epsilon^*} \phi \left(\frac{x}{\sigma_\epsilon^*} \right) \int_{-\infty}^{\infty} \mathrm{e}^{-\frac{1}{2}z^2} \Phi \left(\frac{x - \sigma^{*2}/2 - \kappa_1 - \sqrt{\kappa_2} z}{\sigma^*} \right) \mathrm{d}z$$

$$= \frac{1}{\sigma_\epsilon^*} \phi \left(\frac{x}{\sigma_\epsilon^*} \right) \Phi \left(\frac{x - \sigma^{*2}/2 - \kappa_1}{\sqrt{\sigma^{*2} + \kappa_2}} \right),$$

where in the second step the substitution $z = (y - \kappa_1)/\sqrt{\kappa_2}$ and in the third step Identity 3 from Appendix B have been used. For the second integral in Equation (3.3) it follows analogously

$$\begin{aligned}
\int_{-\infty}^{\infty} p_{\Xi^*}^{(t)}(x) p_{\xi^*}(y) \left[1 - P_{\xi^*|\Xi^*}^{(acc)}(y|x)\right] \mathrm{d}y \\
&= \frac{1}{\sigma_\epsilon^* \sqrt{\kappa_2}} \int_{-\infty}^{\infty} \phi\left(\frac{x-\kappa_1}{\sqrt{\kappa_2}}\right) \phi\left(\frac{y}{\sigma_\epsilon^*}\right) \left[1 - \Phi\left(\frac{y - x - \sigma^{*2}/2}{\sigma^*}\right)\right] \mathrm{d}y \\
&= \frac{1}{\sqrt{2\pi\kappa_2}} \phi\left(\frac{x-\kappa_1}{\sqrt{\kappa_2}}\right) \int_{-\infty}^{\infty} \mathrm{e}^{-\frac{1}{2}z^2} \left[1 - \Phi\left(\frac{\sigma_\epsilon^* z - x - \sigma^{*2}/2}{\sigma^*}\right)\right] \mathrm{d}z \\
&= \frac{1}{\sqrt{\kappa_2}} \phi\left(\frac{x-\kappa_1}{\sqrt{\kappa_2}}\right) \Phi\left(\frac{x + \sigma^{*2}/2}{\sqrt{\sigma^{*2} + \sigma_\epsilon^{*2}}}\right),
\end{aligned}$$

where in the second step the substitution $z = y/\sigma_\epsilon^*$ and in the third step Identity 3 from Appendix B have been used. Putting together the results it follows

$$\begin{aligned}
p_{\Xi^*}^{(t+1)}(x) = \frac{1}{\sigma_\epsilon^*} \phi\left(\frac{x}{\sigma_\epsilon^*}\right) \Phi\left(\frac{x - \sigma^{*2}/2 - \kappa_1}{\sqrt{\sigma^{*2} + \kappa_2}}\right) \\
+ \frac{1}{\sqrt{\kappa_2}} \phi\left(\frac{x-\kappa_1}{\sqrt{\kappa_2}}\right) \Phi\left(\frac{x + \sigma^{*2}/2}{\sqrt{\sigma^{*2} + \sigma_\epsilon^{*2}}}\right) \qquad \text{(C.4)}
\end{aligned}$$

for the probability density of $\Xi^{*(t+1)}$.

2.2 Computing the Mean

With the probability density $p_{\Xi^*}^{(t+1)}$ expressed as the sum of two terms as in Equation (C.4), computation of the mean

$$\mathrm{E}\left[\Xi^{*(t+1)}\right] = \int_{-\infty}^{\infty} x p_{\Xi^*}^{(t+1)}(x) \mathrm{d}x$$

involves the evaluation of two integrals. Evaluating the first of these yields

$$\begin{aligned}
\frac{1}{\sigma_\epsilon^*} \int_{-\infty}^{\infty} x \phi\left(\frac{x}{\sigma_\epsilon^*}\right) \Phi\left(\frac{x - \sigma^{*2}/2 - \kappa_1}{\sqrt{\sigma^{*2} + \kappa_2}}\right) \mathrm{d}x \\
&= \frac{\sigma_\epsilon^*}{\sqrt{2\pi}} \int_{-\infty}^{\infty} y \mathrm{e}^{-\frac{1}{2}y^2} \Phi\left(\frac{\sigma_\epsilon^* y - \sigma^{*2}/2 - \kappa_1}{\sqrt{\sigma^{*2} + \kappa_2}}\right) \mathrm{d}y \\
&= \frac{\sigma_\epsilon^{*2}}{\sqrt{\sigma^{*2} + \sigma_\epsilon^{*2} + \kappa_2}} \phi\left(\frac{\sigma^{*2}/2 + \kappa_1}{\sqrt{\sigma^{*2} + \sigma_\epsilon^{*2} + \kappa_2}}\right) \\
&= \sigma_\epsilon^{*2} \chi^{(1)},
\end{aligned}$$

where in the first step the substitution $y = x/\sigma_\epsilon^*$ and in the second step Identity 3 from Appendix B have been used. The final step follows from comparison with Equation (C.1). Analogously, evaluation of the second integral yields

$$\begin{aligned}
\frac{1}{\sqrt{\kappa_2}} \int_{-\infty}^{\infty} x \phi\left(\frac{x-\kappa_1}{\sqrt{\kappa_2}}\right) \Phi\left(\frac{x + \sigma^{*2}/2}{\sqrt{\sigma^{*2} + \sigma_\epsilon^{*2}}}\right) \mathrm{d}x \\
&= \frac{1}{\sqrt{2\pi}} \int_{-\infty}^{\infty} (\kappa_1 + \sqrt{\kappa_2} y) \mathrm{e}^{-\frac{1}{2}y^2} \Phi\left(\frac{\sqrt{\kappa_2} y + \sigma^{*2}/2 + \kappa_1}{\sqrt{\sigma^{*2} + \sigma_\epsilon^{*2}}}\right) \mathrm{d}y
\end{aligned}$$

$$
\begin{aligned}
&= \kappa_1 \Phi\left(\frac{\sigma^{*2}/2+\kappa_1}{\sqrt{\sigma^{*2}+\sigma_\epsilon^{*2}+\kappa_2}}\right) + \frac{\kappa_2}{\sqrt{\sigma^{*2}+\sigma_\epsilon^{*2}+\kappa_2}}\phi\left(\frac{\sigma^{*2}/2+\kappa_1}{\sqrt{\sigma^{*2}+\sigma_\epsilon^{*2}+\kappa_2}}\right)\\
&= \kappa_1\chi^{(0)} + \kappa_2\chi^{(1)},
\end{aligned}
$$

where in the first step the substitution $y = (x-\kappa_1)/\sqrt{\kappa_2}$ and in the second step Identity 3 from Appendix B have been used. Again, the final step follows from comparison with Equation (C.1). Altogether, it follows

$$
\mathrm{E}\left[\Xi^{*(t+1)}\right] = \kappa_1\chi^{(0)} + \left(\sigma_\epsilon^{*2}+\kappa_2\right)\chi^{(1)}. \tag{C.5}
$$

for the mean of $\Xi^{*(t+1)}$.

2.3 Computing the Variance

With the probability density $p_{\Xi^*}^{(t+1)}$ expressed as the sum of two terms as in Equation (C.4), computation of the variance

$$
\mathrm{Var}\left[\Xi^{*(t+1)}\right] = \int_{-\infty}^{\infty} x^2 p_{\Xi^*}^{(t+1)}(x)\mathrm{d}x - \left(\mathrm{E}\left[\Xi^{*(t+1)}\right]\right)^2
$$

involves the evaluation of two integrals. Evaluating the first of these yields

$$
\begin{aligned}
&\frac{1}{\sigma_\epsilon^*}\int_{-\infty}^{\infty} x^2\phi\left(\frac{x}{\sigma_\epsilon^*}\right)\Phi\left(\frac{x-\sigma^{*2}/2-\kappa_1}{\sqrt{\sigma^{*2}+\kappa_2}}\right)\mathrm{d}x\\
&= \frac{\sigma_\epsilon^{*2}}{\sqrt{2\pi}}\int_{-\infty}^{\infty}\left(1+\left(y^2-1\right)\right)\mathrm{e}^{-\frac{1}{2}y^2}\Phi\left(\frac{\sigma_\epsilon^* y-\sigma^{*2}/2-\kappa_1}{\sqrt{\sigma^{*2}+\kappa_2}}\right)\mathrm{d}y\\
&= \sigma_\epsilon^{*2}\left[1-\Phi\left(\frac{\sigma^{*2}/2+\kappa_1}{\sqrt{\sigma^{*2}+\sigma_\epsilon^{*2}+\kappa_2}}\right)\right] - \frac{\sigma_\epsilon^{*4}\left(-\sigma^{*2}/2-\kappa_1\right)}{\left(\sigma^{*2}+\sigma_\epsilon^{*2}+\kappa_2\right)^{3/2}}\phi\left(\frac{\sigma^{*2}/2+\kappa_1}{\sqrt{\sigma^{*2}+\sigma_\epsilon^{*2}+\kappa_2}}\right)\\
&= \sigma_\epsilon^{*2}\left(1-\chi^{(0)}\right) - \sigma_\epsilon^{*4}\chi^{(2)},
\end{aligned}
$$

where substitution $y = x/\sigma_\epsilon^*$ has been used in the first step and Identity 3 from Appendix B in the second. The final step follows from comparison with Equation (C.1). Analogously, evaluation of the second integral yields

$$
\begin{aligned}
&\frac{1}{\sqrt{\kappa_2}}\int_{-\infty}^{\infty} x^2\phi\left(\frac{x-\kappa_1}{\sqrt{\kappa_2}}\right)\Phi\left(\frac{x+\sigma^{*2}/2}{\sqrt{\sigma^{*2}+\sigma_\epsilon^{*2}}}\right)\mathrm{d}x\\
&= \int_{-\infty}^{\infty}\left(\left(\kappa_1^2+\kappa_2\right)+2\kappa_1\sqrt{\kappa_2}y+\kappa_2\left(y^2-1\right)\right)\phi(y)\Phi\left(\frac{\sqrt{\kappa_2}y+\sigma^{*2}/2+\kappa_1}{\sqrt{\sigma^{*2}+\sigma_\epsilon^{*2}}}\right)\mathrm{d}y\\
&= \left(\kappa_1^2+\kappa_2\right)\Phi\left(\frac{\sigma^{*2}/2+\kappa_1}{\sqrt{\sigma^{*2}+\sigma_\epsilon^{*2}+\kappa_2}}\right) + \frac{2\kappa_1\kappa_2}{\sqrt{\sigma^{*2}+\sigma_\epsilon^2+\kappa_2}}\phi\left(\frac{\sigma^{*2}/2+\kappa_1}{\sqrt{\sigma^{*2}+\sigma_\epsilon^2+\kappa_2}}\right)\\
&\qquad - \frac{\kappa_2^2\left(\sigma^{*2}/2+\kappa_1\right)}{\left(\sigma^{*2}+\sigma_\epsilon^{*2}+\kappa_2\right)^{3/2}}\phi\left(\frac{\sigma^{*2}/2+\kappa_1}{\sqrt{\sigma^{*2}+\sigma_\epsilon^2+\kappa_2}}\right)\\
&= \left(\kappa_1^2+\kappa_2\right)\chi^{(0)} + 2\kappa_1\kappa_2\chi^{(1)} + \kappa_2^2\chi^{(2)},
\end{aligned}
$$

where the substitution $y = (x - \kappa_1)/\sqrt{\kappa_2}$ has been used in the first step and Identity 3 from Appendix B in the second. The final step follows again from comparison with Equation (C.1). Altogether, making use of Equations (C.1) and (C.5) it follows

$$\begin{aligned}
\mathrm{Var}\left[\Xi^{*\,(t+1)}\right] &= \sigma_\epsilon^{*\,2} + \left(\kappa_1^2 + \kappa_2 - \sigma_\epsilon^{*\,2}\right)\chi^{(0)} + 2\kappa_1\kappa_2\chi^{(1)} \\
&\qquad - \left(\sigma_\epsilon^{*\,4} - \kappa_2^2\right)\chi^{(2)} - \left(\mathrm{E}\left[\Xi^{*\,(t+1)}\right]\right)^2 \\
&= \kappa_2 + \kappa_1\left[\kappa_1\chi^{(0)} + \left(\sigma_\epsilon^{*\,2} + \kappa_2\right)\chi^{(1)}\right] - \left(\mathrm{E}\left[\Xi^{*\,(t+1)}\right]\right)^2 \\
&\qquad + \left(\sigma_\epsilon^{*\,2} - \kappa_2\right)\left[1 - \chi^{(0)} - \kappa_1\chi^{(1)} - \left(\sigma_\epsilon^{*\,2} + \kappa_2\right)\chi^{(2)}\right] \\
&= \kappa_2 + \kappa_1\mathrm{E}\left[\Xi^{*\,(t+1)}\right] - \left(\mathrm{E}\left[\Xi^{*\,(t+1)}\right]\right)^2 \\
&\qquad + \left(\sigma_\epsilon^{*\,2} - \kappa_2\right)\left(1 - \frac{\partial}{\partial\kappa_1}\mathrm{E}\left[\Xi^{*\,(t+1)}\right]\right)
\end{aligned} \tag{C.6}$$

for the variance of $\Xi^{*\,(t+1)}$.

3. Determining the Stability of the Fixed Point

With g_1 and g_2 given by Equations (3.7) and (3.8), respectively, the entries of the Jacobian defined in Equation (3.13) at the fixed point are

$$\begin{aligned}
m_{11} &= \frac{\partial g_1}{\partial \kappa_1} = \chi^{(0)} + \kappa_1\chi^{(1)} + 2\kappa_2\chi^{(2)} \\
m_{12} &= \frac{\partial g_1}{\partial \kappa_2} = \chi^{(1)} + \frac{1}{2}\kappa_1\chi^{(2)} + \kappa_2\chi^{(3)} \\
m_{21} &= \frac{\partial g_2}{\partial \kappa_1} = g_1 + (\kappa_1 - 2g_1)\frac{\partial g_1}{\partial \kappa_1} - \left(\sigma_\epsilon^{*\,2} - \kappa_2\right)\frac{\partial^2 g_1}{\partial \kappa_1^2} = (1 - m_{11})\kappa_1 \\
m_{22} &= \frac{\partial g_2}{\partial \kappa_2} = (\kappa_1 - 2g_1)\frac{\partial g_1}{\partial \kappa_2} + \frac{\partial g_1}{\partial \kappa_1} - \left(\sigma_\epsilon^{*\,2} - \kappa_2\right)\frac{\partial^2 g_1}{\partial \kappa_1 \partial \kappa_2} = m_{11} - m_{12}\kappa_1,
\end{aligned}$$

where Equations (C.1), (C.2), (3.10), and (3.11) have been used. Stability of the iterated map is determined by the eigenvalues

$$\eta_{1,2} = \frac{m_{11} + m_{22}}{2} \pm \sqrt{\left(\frac{m_{11} - m_{22}}{2}\right)^2 + m_{12}m_{21}} \tag{C.7}$$

of the Jacobian.

4. Success Probability and Quality Gain

Success probability and quality gain can be obtained easily by solving the integrals in Equations (3.14) and (3.16), respectively.

4.1 Determination of the Success Probability

The success probability P_{succ} is given by Equation (3.14). Exchanging the order of the integrations,

$$P_{succ} = \int_{-\infty}^{\infty} p_{\xi^*}(y) \int_{-\infty}^{\infty} p_{\Xi^*}(x) P^{(acc)}_{\xi^*|\Xi^*}(y|x)\mathrm{d}x\mathrm{d}y.$$

Using Equations (3.1) and (3.2), it follows for the inner integral

$$\begin{aligned}\int_{-\infty}^{\infty} p_{\Xi^*}(x) P^{(acc)}_{\xi^*|\Xi^*}(y|x)\mathrm{d}x &= \frac{1}{\sqrt{\kappa_2}} \int_{-\infty}^{\infty} \phi\left(\frac{x-\kappa_1}{\sqrt{\kappa_2}}\right) \Phi\left(\frac{y-x-{\sigma^*}^2/2}{\sigma^*}\right) \mathrm{d}x \\ &= \frac{1}{\sqrt{2\pi}} \int_{-\infty}^{\infty} \mathrm{e}^{-\frac{1}{2}z^2} \Phi\left(\frac{y-\kappa_1-\sqrt{\kappa_2}z-{\sigma^*}^2/2}{\sigma^*}\right) \mathrm{d}z \\ &= \Phi\left(\frac{y-{\sigma^*}^2/2-\kappa_1}{\sqrt{{\sigma^*}^2+\kappa_2}}\right),\end{aligned}$$

where the substitution $z = (x-\kappa_1)/\sqrt{\kappa_2}$ has been used in the second step and Identity 3 from Appendix B in the third. Therefore, using the fact that ξ^* is normally distributed with mean zero and with variance ${\sigma_\epsilon^*}^2$,

$$\begin{aligned}P_{succ} &= \frac{1}{\sigma_\epsilon^*} \int_{-\infty}^{\infty} \phi\left(\frac{y}{\sigma_\epsilon^*}\right) \Phi\left(\frac{y-{\sigma^*}^2/2-\kappa_1}{\sqrt{{\sigma^*}^2+\kappa_2}}\right) \mathrm{d}y \\ &= \frac{1}{\sqrt{2\pi}} \int_{-\infty}^{\infty} \mathrm{e}^{-\frac{1}{2}z^2} \Phi\left(\frac{\sigma_\epsilon^* z-{\sigma^*}^2/2-\kappa_1}{\sqrt{{\sigma^*}^2+\kappa_2}}\right) \mathrm{d}z \\ &= 1-\Phi\left(\frac{{\sigma^*}^2/2+\kappa_1}{\sqrt{{\sigma^*}^2+{\sigma_\epsilon^*}^2+\kappa_2}},\right) \\ &= 1-\chi^{(0)}, \qquad \text{(C.8)}\end{aligned}$$

where the substitution $z = y/\sigma_\epsilon^*$ has been used in the first step and Identity 3 from Appendix B in the second. The final step follows from comparison with Equation (C.1).

4.2 Determination of the Quality Gain

The quality gain is given by Equation (3.16). Exchanging the order of the integrations,

$$\mathrm{E}\left[\Delta^*_{sphere}\right] = \int_{-\infty}^{\infty} y p_{\delta^*}(y) \int_{-\infty}^{\infty} p_{\Xi^*}(x) P^{(acc)}_{\delta^*|\Xi^*}(y|x)\mathrm{d}x\mathrm{d}y.$$

Using Equations (3.1) and (3.17), it follows for the inner integral

$$\begin{aligned}\int_{-\infty}^{\infty} p_{\Xi^*}(x) P^{(acc)}_{\delta^*|\Xi^*}(y|x)\mathrm{d}x &= \frac{1}{\sqrt{\kappa_2}} \int_{-\infty}^{\infty} \phi\left(\frac{x-\kappa_1}{\sqrt{\kappa_2}}\right) \Phi\left(\frac{y-x}{\sigma_\epsilon^*}\right) \mathrm{d}x \\ &= \frac{1}{\sqrt{2\pi}} \int_{-\infty}^{\infty} \mathrm{e}^{-\frac{1}{2}z^2} \Phi\left(\frac{y-\kappa_1-\sqrt{\kappa_2}z}{\sigma_\epsilon^*}\right) \mathrm{d}z \\ &= \Phi\left(\frac{y-\kappa_1}{\sqrt{{\sigma_\epsilon^*}^2+\kappa_2}}\right),\end{aligned}$$

where in the second step the substitution $z = (x-\kappa_1)/\sqrt{\kappa_2}$ and in the third step Identity 3 from Appendix B have been used. Therefore, using the fact that δ^* is normally distributed with mean $-{\sigma^*}^2/2$ and with variance ${\sigma^*}^2$,

$$\begin{aligned}\mathrm{E}\left[\Delta^*_{sphere}\right] &= \frac{1}{\sigma^*} \int_{-\infty}^{\infty} y\phi\left(\frac{y+{\sigma^*}^2/2}{\sigma^*}\right) \Phi\left(\frac{y-\kappa_1}{\sqrt{{\sigma_\epsilon^*}^2+\kappa_2}}\right) \mathrm{d}y \\ &= \frac{1}{\sqrt{2\pi}} \int_{-\infty}^{\infty} \left(-\frac{{\sigma^*}^2}{2}+\sigma^* z\right) \mathrm{e}^{-\frac{1}{2}z^2} \Phi\left(\frac{-{\sigma^*}^2/2+\sigma^* z-\kappa_1}{\sqrt{{\sigma_\epsilon^*}^2+\kappa_2}}\right) \mathrm{d}z\end{aligned}$$

$$
\begin{aligned}
&= -\frac{\sigma^{*2}}{2}\left[1-\Phi\left(\frac{\sigma^{*2}/2+\kappa_1}{\sqrt{\sigma^{*2}+\sigma_\epsilon^{*2}+\kappa_2}}\right)\right] \\
&\qquad\qquad + \frac{\sigma^{*2}}{\sqrt{\sigma^{*2}+\sigma_\epsilon^{*2}+\kappa_2}}\phi\left(\frac{\sigma^{*2}/2+\kappa_1}{\sqrt{\sigma^{*2}+\sigma_\epsilon^{*2}+\kappa_2}}\right) \\
&= -\frac{\sigma^{*2}}{2}\left(1-\chi^{(0)}\right)+\sigma^{*2}\chi^{(1)}, \qquad \text{(C.9)}
\end{aligned}
$$

where in the first step substitution $z = (y - \kappa_1)/\sqrt{\kappa_2}$ and in the second step Identity 3 from Appendix B have been used. The final step follows again from comparison with Equation (C.1).

Appendix D
Determining the Effects of Sampling and Selection

The purpose of this appendix is to determine the effects of sampling from a distribution and of subsequent selection of the largest observed values based on noisy information. The approach pursued is moment-based and assumes that the distribution to be sampled from can be characterized by its first few moments. The problem to be analyzed can be stated concisely as follows:

> *Suppose that $x_1, \ldots, x_\lambda$ are λ jointly distributed, independent random variables with distribution function P and with moments μ'_j, $j \geq 1$. Associated with every x_k is what we refer to as a measured value that is drawn from a normal distribution with mean x_k and with variance ϑ^2. For $\mu \leq \lambda$, the expected moments of the sample consisting of those μ of the x_k that have the largest associated measured values are to be computed.*

The sample consisting of those μ of the x_k that have the largest associated measured values will be called the *truncated sample*. We denote the x_k with the ith largest associated measured value by $x_{i;\lambda}$ and refer to it as the ith *noisy order statistic*. As ordinary order statistics, noisy order statistics have the distinctive property that they are no longer independent nor identically distributed even though the original observations were so. Only in the limit of infinite noise strength or if $\mu = \lambda$, selection is entirely random and independence of the members of the truncated sample is regained.

The organization of this appendix is as follows. In Section 1, central moments of the truncated sample are represented as sums of products of the $x_{i;\lambda}$. In Section 2, expected moments of the truncated sample in the limit of infinite noise strength, i.e. under trivial selection, are computed. These results have been derived previously for example by Kenney and Keeping [51]. For finite ϑ, the situation is considerably more complicated. In Section 3, joint probability density functions of noisy order statistics are derived, and integral representations for the expected values of sums of products of noisy order statistics are found. In Section 4, the probability density function of the distribution from which the x_i are drawn is expanded in terms of derivatives of the normal distribution function, making it possible to solve the integrals in the previously obtained representation and to express the expected values of sums of products of noisy order statistics in terms of coefficients that can be obtained numerically. Finally, results from Sections 1 and 4 are combined to obtain expected moments of the truncated sample. A simple corollary for normal samples that is used in Chapter 5 can be found in Section 5. Section 6 contains a *Mathematica* program handling the tedious details of the calculations in Section 4.

1. Sample Moments as Sums of Products

As a first step towards computing expected central moments of the truncated sample, let us express the moments as sums of products of the $x_{i;\lambda}$. The mean of the truncated sample is denoted as

$$\langle x \rangle = m_1 = \frac{1}{\mu} \sum_{i=1}^{\mu} x_{i;\lambda}. \tag{D.1}$$

For notational brevity, the ranges of sums are omitted in much of what follows. Unless noted otherwise, sums range from 1 to μ for all summation indices. Moreover, whenever the ranges are omitted, we follow the convention that multiple sums over products of variables exclude equal indices. For example, we have

$$\sum x_i x_j = x_1 x_2 + x_1 x_3 + x_2 x_3$$
$$\sum x_i^2 x_j = x_1^2 x_2 + x_1^2 x_3 + x_1 x_2^2 + x_2^2 x_3 + x_1 x_3^2 + x_2 x_3^2$$

for sums extending over a range of size three.

The Sample Variance

According to Equation (4.1),

$$\begin{aligned} m_2 &= \frac{1}{\mu} \sum_{i=1}^{\mu} \left(x_{i;\lambda} - \langle x \rangle \right)^2 \\ &= \frac{1}{\mu} \sum x_{i;\lambda}^2 - \left(\frac{1}{\mu} \sum x_{i;\lambda} \right)^2 . \end{aligned}$$

Using the identity

$$\left(\sum x_i \right)^2 = \sum x_i^2 + 2 \sum x_i x_j$$

it follows

$$m_2 = \frac{\mu - 1}{\mu^2} \sum x_{i;\lambda}^2 - \frac{2}{\mu^2} \sum x_{i;\lambda} x_{j;\lambda} \tag{D.2}$$

for the variance of the truncated sample.

The Third Central Sample Moment

According to Equation (4.1),

$$\begin{aligned} m_3 &= \frac{1}{\mu} \sum_{i=1}^{\mu} \left(x_{i;\lambda} - \langle x \rangle \right)^3 \\ &= \frac{1}{\mu} \sum x_{i;\lambda}^3 - 3 \left(\frac{1}{\mu} \sum x_{i;\lambda}^2 \right) \left(\frac{1}{\mu} \sum x_{j;\lambda} \right) + 2 \left(\frac{1}{\mu} \sum x_{i;\lambda} \right)^3 . \end{aligned}$$

Using the identities

$$\left(\sum x_i^2 \right) \left(\sum x_j \right) = \sum x_i^3 + \sum x_i^2 x_j$$
$$\left(\sum x_i \right)^3 = \sum x_i^3 + 3 \sum x_i^2 x_j + 6 \sum x_i x_j x_k$$

it follows

$$m_3 = \frac{\mu^2 - 3\mu + 2}{\mu^3} \sum x_{i;\lambda}^3 - \frac{3\mu - 6}{\mu^3} \sum x_{i;\lambda}^2 x_{j;\lambda} + \frac{12}{\mu^3} \sum x_{i;\lambda} x_{j;\lambda} x_{k;\lambda} \tag{D.3}$$

for the third central moment of the truncated sample.

The Fourth Central Sample Moment

According to Equation (4.1),

$$m_4 = \frac{1}{\mu}\sum_{i=1}^{\mu}(x_{i;\lambda} - \langle x\rangle)^4$$
$$= \frac{1}{\mu}\sum x_{i;\lambda}^4 - 4\left(\frac{1}{\mu}\sum x_{i;\lambda}^3\right)\left(\frac{1}{\mu}\sum x_{j;\lambda}\right) + 6\left(\frac{1}{\mu}\sum x_{i;\lambda}^2\right)\left(\frac{1}{\mu}\sum x_{j;\lambda}\right)^2 - 3\left(\frac{1}{\mu}\sum x_{i;\lambda}\right)^4.$$

Using the identities

$$\left(\sum x_i^3\right)\left(\sum x_j\right) = \sum x_i^4 + \sum x_i^3 x_j$$
$$\left(\sum x_i^2\right)\left(\sum x_j\right)^2 = \sum x_i^4 + 2\sum x_i^3 x_j + 2\sum x_i^2 x_j^2 + 2\sum x_i^2 x_j x_k$$
$$\left(\sum x_i\right)^4 = \sum x_i^4 + 4\sum x_i^3 x_j + 6\sum x_i^2 x_j^2 + 12\sum x_i^2 x_j x_k + 24\sum x_i x_j x_k x_l$$

it follows

$$m_4 = \frac{\mu^3 - 4\mu^2 + 6\mu - 3}{\mu^4}\sum x_{i;\lambda}^4 - \frac{4\mu^2 - 12\mu + 12}{\mu^4}\sum x_{i;\lambda}^3 x_{j;\lambda} + \frac{12\mu - 18}{\mu^4}\sum x_{i;\lambda}^2 x_{j;\lambda}^2 + \frac{12\mu - 36}{\mu^4}\sum x_{i;\lambda}^2 x_{j;\lambda} x_{k;\lambda} - \frac{72}{\mu^4}\sum x_{i;\lambda} x_{j;\lambda} x_{k;\lambda} x_{l;\lambda} \quad \text{(D.4)}$$

for the fourth central moment of the truncated sample.

The Square of the Sample Variance

According to Equation (4.1),

$$m_2^2 = \left(\frac{1}{\mu}\sum_{i=1}^{\mu}(x_{i;\lambda} - \langle x\rangle)^2\right)^2$$
$$= \left(\frac{1}{\mu}\sum x_{i;\lambda}^2\right)^2 - 2\left(\frac{1}{\mu}\sum x_{i;\lambda}^2\right)\left(\frac{1}{\mu}\sum x_{j;\lambda}\right)^2 + \left(\frac{1}{\mu}\sum x_{i;\lambda}\right)^4.$$

Using the identities

$$\left(\sum x_i^2\right)^2 = \sum x_i^4 + 2\sum x_i^2 x_j^2$$
$$\left(\sum x_i^2\right)\left(\sum x_j\right)^2 = \sum x_i^4 + 2\sum x_i^3 x_j + 2\sum x_i^2 x_j^2 + 2\sum x_i^2 x_j x_k$$
$$\left(\sum x_i\right)^4 = \sum x_i^4 + 4\sum x_i^3 x_j + 6\sum x_i^2 x_j^2 + 12\sum x_i^2 x_j x_k + 24\sum x_i x_j x_k x_l$$

it follows

$$m_2^2 = \frac{\mu^2 - 2\mu + 1}{\mu^4}\sum x_{i;\lambda}^4 - \frac{4\mu - 4}{\mu^4}\sum x_{i;\lambda}^3 x_{j;\lambda} + \frac{2\mu^2 - 4\mu + 6}{\mu^4}\sum x_{i;\lambda}^2 x_{j;\lambda}^2 - \frac{4\mu - 12}{\mu^4}\sum x_{i;\lambda}^2 x_{j;\lambda} x_{k;\lambda} + \frac{24}{\mu^4}\sum x_{i;\lambda} x_{j;\lambda} x_{k;\lambda} x_{l;\lambda} \quad \text{(D.5)}$$

for the square of the variance of the truncated sample.

2. The Infinite Noise Limit

In the limit of infinite noise strength, computing expected values of the terms on the right hand sides of Equations (D.2), (D.3), (D.4), and (D.5) is easy as the random variables that occur are independent. The problem at hand reduces to computing sampling moments. The expected values of products of powers of the $x_{i;\lambda}$ can be expressed simply as products of moments of the distribution from which the x_k are drawn. The results presented in this section have been given before by Kenney and Keeping [51].

The Sampling Variance

The first sum in Equation (D.2) contains μ independent terms, each with expected value μ_2'. The second sum contains $\mu(\mu-1)/2$ independent terms, each with expected value ${\mu_1'}^2$. Therefore,

$$\mathrm{E}\left[\sum x_{i;\lambda}^2\right] = \mu\mu_2'$$

$$\mathrm{E}\left[\sum x_{i;\lambda}x_{j;\lambda}\right] = \frac{\mu(\mu-1)}{2}{\mu_1'}^2.$$

It follows that the expected sample variance is

$$\begin{aligned}\mathrm{E}\left[m_2\right] &= \frac{\mu-1}{\mu}\left(\mu_2' - {\mu_1'}^2\right)\\ &= \frac{\mu-1}{\mu}\mu_2,\end{aligned} \tag{D.6}$$

where Equations (A.4) have been used to replace moments about zero by central moments.

The Third Central Sampling Moment

The first sum in Equation (D.3) contains μ independent terms each with expected value μ_3'. The second sum contains $\mu(\mu-1)$ terms, each with expected value $\mu_2'\mu_1'$. The third sum contains $\mu(\mu-1)(\mu-2)/6$ independent terms, each with expected value ${\mu_1'}^3$. Therefore,

$$\mathrm{E}\left[\sum x_{i;\lambda}^3\right] = \mu\mu_3'$$

$$\mathrm{E}\left[\sum x_{i;\lambda}^2 x_{j;\lambda}\right] = \mu(\mu-1)\mu_2'\mu_1'$$

$$\mathrm{E}\left[\sum x_{i;\lambda}x_{j;\lambda}x_{k;\lambda}\right] = \frac{\mu(\mu-1)(\mu-2)}{6}{\mu_1'}^3.$$

It follows that the expected third central moment of the sample is

$$\begin{aligned}\mathrm{E}\left[m_3\right] &= \frac{(\mu-1)(\mu-2)}{\mu^2}\left(\mu_3' - 3\mu_2'\mu_1' + 2{\mu_1'}^3\right)\\ &= \frac{(\mu-1)(\mu-2)}{\mu^2}\mu_3,\end{aligned} \tag{D.7}$$

where Equations (A.4) have been used to replace moments about zero by central moments.

The Fourth Central Sampling Moment and the Expected Square of the Sample Variance

The first sum in Equations (D.4) and (D.5) contains μ independent terms, each with expected value μ_4'. The second sum contains $\mu(\mu-1)$ independent terms, each with expected value $\mu_3'\mu_1'$.

The third sum contains $\mu(\mu-1)/2$ independent terms, each with expected value ${\mu'_2}^2$. The fourth sum contains $\mu(\mu-1)(\mu-2)/2$ independent terms, each with expected value $\mu'_2{\mu'_1}^2$. The fifth sum contains $\mu(\mu-1)(\mu-2)(\mu-3)/24$ independent terms, each with expected value ${\mu'_1}^4$. Therefore,

$$\begin{aligned}
\mathrm{E}\left[\sum x_{i;\lambda}^4\right] &= \mu\mu'_4 \\
\mathrm{E}\left[\sum x_{i;\lambda}^3 x_{j;\lambda}\right] &= \mu(\mu-1)\mu'_3\mu'_1 \\
\mathrm{E}\left[\sum x_{i;\lambda}^2 x_{j;\lambda}^2\right] &= \frac{\mu(\mu-1)}{2}{\mu'_2}^2 \\
\mathrm{E}\left[\sum x_{i;\lambda}^2 x_{j;\lambda} x_{k;\lambda}\right] &= \frac{\mu(\mu-1)(\mu-2)}{2}\mu'_2{\mu'_1}^2 \\
\mathrm{E}\left[\sum x_{i;\lambda} x_{j;\lambda} x_{k;\lambda} x_{l;\lambda}\right] &= \frac{\mu(\mu-1)(\mu-2)(\mu-3)}{24}{\mu'_1}^4 .
\end{aligned}$$

It follows that the expected fourth central moment of the sample is

$$\begin{aligned}
\mathrm{E}\left[m_4\right] &= \frac{(\mu-1)(\mu^2-3\mu+3)}{\mu^3}\mu'_4 - \frac{4(\mu-1)(\mu^2-3\mu+3)}{\mu^3}\mu'_3\mu'_1 \\
&\quad + \frac{3(\mu-1)(2\mu-3)}{\mu^3}{\mu'_2}^2 + \frac{6(\mu-1)(\mu-2)(\mu-3)}{\mu^3}\mu'_2{\mu'_1}^2 \\
&\quad - \frac{3(\mu-1)(\mu-2)(\mu-3)}{\mu^3}{\mu'_1}^4 \\
&= \frac{(\mu-1)(\mu^2-3\mu+3)}{\mu^3}\mu_4 + \frac{3(\mu-1)(2\mu-3)}{\mu^3}\mu_2^2,
\end{aligned} \tag{D.8}$$

and that the expected squared sample variance is

$$\begin{aligned}
\mathrm{E}\left[m_2^2\right] &= \frac{(\mu-1)^2}{\mu^3}\mu'_4 - \frac{4(\mu-1)^2}{\mu^3}\mu'_3\mu'_1 + \frac{(\mu-1)(\mu^2-2\mu+3)}{\mu^3}{\mu'_2}^2 \\
&\quad - \frac{2(\mu-1)(\mu-2)(\mu-3)}{\mu^3}\mu'_2{\mu'_1}^2 + \frac{(\mu-1)(\mu-2)(\mu-3)}{\mu^3}{\mu'_1}^4 \\
&= \frac{(\mu-1)^2}{\mu^3}\mu_4 + \frac{(\mu-1)(\mu^2-2\mu+3)}{\mu^3}\mu_2^2,
\end{aligned} \tag{D.9}$$

where Equations (A.4) have been used to replace moments about zero by central moments.

3. Noisy Order Statistics

If selection is nontrivial, computing expected values of the sums in Equations (D.2), (D.3), (D.4), and (D.5) is not as easy. In the absence of noise, joint moments of the first μ ordinary order statistics need to be determined. If there is noise present, the ordering of the sample members is based on measured values and does not necessarily reflect the members' ideal values. In Section 3.1, joint probability density functions of noisy order statistics are given. In Section 3.2, an integral representation for the expected values of sums of products of noisy order statistics is derived.

3.1 Probability Density Functions of Noisy Order Statistics

Let Q denote the distribution function of the measured values of the sample members. According to the problem statement at the beginning of this section, Q is the convolution of the distribution P from which the sample members are drawn and a normal distribution with mean zero and with variance ϑ^2. Consider one particular sample member. The probability of that sample member being in the range $u \pm \frac{1}{2}\mathrm{d}u$ is $p(u)\mathrm{d}u$. Given that the value of the sample member is u, its measured value v is normally distributed with mean u and with variance ϑ^2. For the sample member to have the ith largest measured value of the sample, $\lambda - i$ members must have smaller measured values and $i-1$ members must larger measured values. As both the values of the sample members and their measured values are generated independently, the probability for that to happen is $[Q(v)]^{\lambda-i}[1-Q(v)]^{i-1}$. Finally, as there are λ possibilities of picking one particular sample member, and there are $i-1$ out of $\lambda-1$ equivalent different possibilities of ordering the remaining sample members, the probability density of the noisy order statistic $x_{i;\lambda}$ is

$$p_{i;\lambda}(u) = \frac{\lambda!}{\vartheta}p(u)\int_{-\infty}^{\infty}\phi\left(\frac{v-u}{\vartheta}\right)\frac{[Q(v)]^{\lambda-i}[1-Q(v)]^{i-1}}{(\lambda-i)!(i-1)!}\mathrm{d}v, \tag{D.10}$$

where ϕ denotes the probability density function of the standardized normal distribution. In the limit of vanishing noise strength, $\phi((v-u)/\vartheta)/\vartheta$ tends to the Delta function $\delta(v-u)$ and the density of the noisy order statistic agrees with that of the ordinary order statistic given for example by Stuart and Ord [84].

Now consider two sample members with values u_1 and u_2. Their measured values v_1 and v_2 are independently normally distributed with variance ϑ^2 and with means u_1 and u_2, respectively. Let $\lambda \geq i_1 > i_2 \geq 1$. For the first of the two sample members to have the i_1th largest measured value and for the second sample member to have the i_2th largest measured value, $\lambda - i_1$ sample members must have measured values less than v_1, $i_1 - i_2 - 1$ sample members must have measured values between v_1 and v_2, and $i_2 - 1$ members must have measured values larger than v_2. The probability for that to happen is $[Q(v_1)]^{\lambda-i_1}[Q(v_2)-Q(v_1)]^{i_1-i_2-1}[1-Q(v_2)]^{i_2-1}$. Overall, there are $\lambda!$ permutations of the $x_{i;\lambda}$. As there are $(\lambda-i_1)!$ permutations of the selected sample members with measured values less than v_1, $(i_1-i_2-1)!$ permutations of the sample members with measured values between v_1 and v_2, and $(i_2-1)!$ permutations of sample members with measured values larger than v_2, it follows that the joint probability density function of the noisy order statistics $x_{i_1;\lambda}$ and $x_{i_2;\lambda}$ is

$$\begin{aligned} p_{i_1,i_2;\lambda}(u_1,u_2) = \frac{\lambda!}{\vartheta^2}p(u_1)p(u_2)\int_{-\infty}^{\infty}\int_{v_1}^{\infty}\phi\left(\frac{v_1-u_1}{\vartheta}\right)\phi\left(\frac{v_2-u_2}{\vartheta}\right) \\ \frac{[Q(v_1)]^{\lambda-i_1}[Q(v_2)-Q(v_1)]^{i_1-i_2-1}[1-Q(v_2)]^{i_2-1}}{(\lambda-i_1)!(i_1-i_2-1)!(i_2-1)!}\mathrm{d}v_2\mathrm{d}v_1. \end{aligned} \tag{D.11}$$

Note that the lower limit of the inner integral is v_1 as $v_2 > v_1$. Again, for $\vartheta \to 0$, the result agrees with the joint probability density of two order statistics in the absence of noise given by Stuart and Ord [84].

As a direct generalization of Equations (D.10) and (D.11), the joint probability density function of the ν noisy order statistics $x_{i_k;\lambda}$, $k = 1,\ldots,\nu$, with $\lambda \geq i_1 > i_2 > \cdots > i_\nu \geq 1$, reads

$$\begin{aligned} p_{i_1,\ldots,i_\nu;\lambda}(u_1,\ldots,u_\nu) = \frac{\lambda!}{\vartheta^\nu}\int_{-\infty}^{\infty}\int_{v_1}^{\infty}\cdots\int_{v_{\nu-1}}^{\infty}\left[\prod_{k=1}^{\nu}p(u_k)\phi\left(\frac{v_k-u_k}{\vartheta}\right)\right] \\ \left[\prod_{k=0}^{\nu}\frac{[Q(v_{k+1})-Q(v_k)]^{i_k-i_{k+1}-1}}{(i_k-i_{k+1}-1)!}\right]\mathrm{d}v_\nu\ldots\mathrm{d}v_2\mathrm{d}v_1, \end{aligned} \tag{D.12}$$

where it is formally assumed that $i_0 = \lambda+1$, $i_{\nu+1} = 0$, $v_0 = -\infty$, and $v_{\nu+1} = \infty$.

3.2 Expected Values of Sums of Products of Noisy Order Statistics

With the joint probability density functions of noisy order statistics so obtained, it is now possible to compute the expected values of the sums of products of noisy order statistics as required for arriving at the expectations of the central moments of the truncated sample. A slight complication arises from the fact that the joint probability density in Equation (D.12) requires that the indices be ordered such that $i_1 > i_2 > \cdots > i_\nu$. So as to restrict the indices in a sum like

$$S_A = \sum x_{i_1;\lambda}^{\alpha_1} \cdots x_{i_\nu;\lambda}^{\alpha_\nu},$$

where $A = (\alpha_1, \ldots, \alpha_\nu)$ is the vector of exponents, let us formally write $\pi_A(u_1, \ldots, u_\nu)$ for the sum of products of powers of the u_i with all permutations of the exponents $\alpha_1, \ldots, \alpha_\nu$. For example,

$$\begin{aligned} \pi_{22}(u_1, u_2) &= u_1^2 u_2^2 \\ \pi_{111}(u_1, u_2, u_3) &= u_1 u_2 u_3 \\ \pi_{211}(u_1, u_2, u_3) &= u_1^2 u_2 u_3 + u_1 u_2^2 u_3 + u_1 u_2 u_3^2. \end{aligned}$$

Then we can write

$$S_A = \sum_{i_1=\nu}^{\mu} \sum_{i_2=\nu-1}^{i_1-1} \cdots \sum_{i_\nu=1}^{i_{\nu-1}-1} \pi_A(x_{i_1;\lambda}, \ldots, x_{i_\nu;\lambda}).$$

The expectation of a sum of products of order statistics with exponents prescribed by A is thus

$$\begin{aligned} \mathrm{E}[S_A] &= \mathrm{E}\left[\sum_{i_1=\nu}^{\mu} \sum_{i_2=\nu-1}^{i_1-1} \cdots \sum_{i_\nu=1}^{i_{\nu-1}-1} \pi_A(x_{i_1;\lambda}, \ldots, x_{i_\nu;\lambda})\right] \\ &= \int_{-\infty}^{\infty} \cdots \int_{-\infty}^{\infty} \sum_{i_1=\nu}^{\mu} \sum_{i_2=\nu-1}^{i_1-1} \cdots \sum_{i_\nu=1}^{i_{\nu-1}-1} \pi_A(u_1, \ldots, u_\nu) \\ &\qquad p_{i_1,\ldots,i_\nu;\lambda}(u_1, \ldots, u_\nu) \mathrm{d}u_\nu \ldots \mathrm{d}u_1. \end{aligned}$$

Using Equation (D.12) to express the joint probability density function of the noisy order statistics and exchanging the order of integrations and summations it follows that

$$\begin{aligned} \mathrm{E}[S_A] = \frac{\lambda!}{\vartheta^\nu} \int_{-\infty}^{\infty} \cdots \int_{-\infty}^{\infty} \pi_A(u_1, \ldots, u_\nu) \\ \int_{-\infty}^{\infty} \int_{v_1}^{\infty} \cdots \int_{v_{\nu-1}}^{\infty} \left[\prod_{k=1}^{\nu} p(u_k) \phi\left(\frac{v_k - u_k}{\vartheta}\right)\right] \\ \sum_{i_1=\nu}^{\mu} \sum_{i_2=\nu-1}^{i_1-1} \cdots \sum_{i_\nu=1}^{i_{\nu-1}-1} \left[\prod_{k=0}^{\nu} \frac{[Q(v_{k+1}) - Q(v_k)]^{i_k - i_{k+1} - 1}}{(i_k - i_{k+1} - 1)!}\right] \\ \mathrm{d}v_\nu \ldots \mathrm{d}v_2 \mathrm{d}v_1 \mathrm{d}u_\nu \ldots \mathrm{d}u_1, \end{aligned}$$

where $i_0 = \lambda + 1$, $i_{\nu+1} = 0$, $v_0 = -\infty$, and $v_{\nu+1} = \infty$. Using Identity 1 from Appendix B it follows that

$$\mathrm{E}[S_A] = \frac{\mu!(\lambda-\mu)}{\vartheta^\nu(\mu-\nu)!}\binom{\lambda}{\mu}\int_{-\infty}^{\infty}\cdots\int_{-\infty}^{\infty}\pi_A(u_1,\ldots,u_\nu)$$
$$\int_{-\infty}^{\infty}\int_{v_1}^{\infty}\cdots\int_{v_{\nu-1}}^{\infty}\left[\prod_{k=1}^{\nu}p(u_k)\phi\left(\frac{v_k-u_k}{\vartheta}\right)\right]$$
$$\int_0^{Q(v_1)} z^{\lambda-\mu-1}[1-z]^{\mu-\nu}\mathrm{d}z\mathrm{d}v_\nu\ldots\mathrm{d}v_2\mathrm{d}v_1\mathrm{d}u_\nu\ldots\mathrm{d}u_1.$$

Finally, substituting $z = \Phi(x)$ and exchanging the order of the integrations yields

$$\mathrm{E}[S_A] = \frac{\mu!}{(\mu-\nu)!}\frac{\lambda-\mu}{\sqrt{2\pi}}\binom{\lambda}{\mu}\int_{-\infty}^{\infty}\mathrm{e}^{-\frac{1}{2}x^2}[\Phi(x)]^{\lambda-\mu-1}[1-\Phi(x)]^{\mu-\nu}I_A(x)\mathrm{d}x \quad \text{(D.13)}$$

with

$$I_A(x) = \frac{1}{\vartheta^\nu}\int_{Q^{-1}(\Phi(x))}^{\infty}\int_{v_1}^{\infty}\cdots\int_{v_{\nu-1}}^{\infty}\int_{-\infty}^{\infty}\cdots\int_{-\infty}^{\infty}\pi_A(u_1,\ldots,u_\nu)$$
$$\left[\prod_{k=1}^{\nu}p(u_k)\phi\left(\frac{v_k-u_k}{\vartheta}\right)\right]\mathrm{d}u_\nu\ldots\mathrm{d}u_1\mathrm{d}v_\nu\ldots\mathrm{d}v_2\mathrm{d}v_1. \quad \text{(D.14)}$$

That is, up to a single remaining integration, the problem of computing $\mathrm{E}[S_A]$ has been reduced to that of computing $I_A(x)$. A moment-based approach for computing $I_A(x)$ is presented in the next section.

4. Expanding the Probability Functions

So as to obtain a closed form expression for $I_A(x)$, let us expand the distribution and probability density functions involved. More specifically, using a Gram-Charlier representation for the probability density p of the distribution from which the sample members are drawn makes it possible to solve the integrals in Equation (D.14). Subsequently expanding $Q^{-1}(\Phi(x))$ into a Cornish-Fisher series yields simple expressions in x. The calculations are not difficult but tedious and lengthy. We had at first done all integrations by hand. The calculations occupied more than sixty pages. Instead of presenting detailed steps here, we merely sketch the derivation and refer to Section 6 where we present a *Mathematica* program that does the calculations. Moreover, to keep intermediate steps and results brief, we neglect all terms involving cumulants higher than the fourth, as well as terms involving products of at least two cumulants higher than the second.

4.1 Solving the Integrals

Rather than characterizing probability density p by means of its central moments μ_j, we choose to switch to using its cumulants κ_j that are related to the moments as reflected by Equation (A.5). So as to solve the integrals in Equation (D.14), let us expand the probability density function p of the distribution from which the x_k are drawn into a Gram-Charlier series. Assuming without loss of generality that the mean of the distribution is zero, according to Equation (A.12) the first terms of the Gram-Charlier series read

$$p(x) = \frac{1}{\sqrt{2\pi\kappa_2}}\exp\left(-\frac{x^2}{2\kappa_2}\right)\left[1+\frac{\gamma_1}{6}\mathrm{He}_3\left(\frac{x}{\sqrt{\kappa_2}}\right)+\frac{\gamma_2}{24}\mathrm{He}_4\left(\frac{x}{\sqrt{\kappa_2}}\right)+\ldots\right], \quad \text{(D.15)}$$

where $\gamma_1 = \kappa_3/\kappa_2^{3/2}$ and $\gamma_2 = \kappa_4/\kappa_2^2$ are the skewness and kurtosis of the distribution, respectively. Introducing new variables $u'_k = u_k/\sqrt{\kappa_2}$ and $v'_k = v_k/\sqrt{\kappa_2+\vartheta^2}$, $k = 1, \ldots, \nu$, in Equation (D.14) yields

$$I_A(x) = \sqrt{\kappa_2}^{\|A\|_1} \sqrt{\kappa_2+\vartheta^2}^{\nu} \int_{v'_0}^{\infty} \cdots \int_{v'_{\nu-1}}^{\infty} J_A(v'_1, \ldots, v'_\nu) \mathrm{d}v'_\nu \ldots \mathrm{d}v'_1, \tag{D.16}$$

where $\|A\|_1 = \sum_{i=1}^{\nu} \alpha_i$, $v'_0 = Q^{-1}(\Phi(x))/\sqrt{\kappa_2+\vartheta^2}$, and

$$J_A(v'_1, \ldots, v'_\nu) = \frac{\sqrt{\kappa_2}^{\nu}}{\vartheta^\nu} \int_{-\infty}^{\infty} \cdots \int_{-\infty}^{\infty} \pi_A(u'_1, \ldots, u'_\nu) \left[\prod_{k=1}^{\nu} p(\sqrt{\kappa_2} u'_k) \phi\left(\frac{\sqrt{\kappa_2}u'_k - \sqrt{\kappa_2+\vartheta^2}v'_k}{\vartheta}\right)\right] \mathrm{d}u'_\nu \ldots \mathrm{d}u'_1. \tag{D.17}$$

Introducing the *noise coefficient*

$$a = \sqrt{\frac{\kappa_2}{\kappa_2+\vartheta^2}}, \tag{D.18}$$

from Identity 2 in Appendix B it follows that

$$\frac{\sqrt{\kappa_2}}{\vartheta} \int_{-\infty}^{\infty} \mathrm{He}_k(u') \mathrm{e}^{-\frac{1}{2}u'^2} \phi\left(\frac{\sqrt{\kappa_2}u' - \sqrt{\kappa_2+\vartheta^2}v'}{\vartheta}\right) \mathrm{d}u' = a^{k+1} \mathrm{He}_k(v') \mathrm{e}^{-\frac{1}{2}v'^2}. \tag{D.19}$$

Therefore, using Equation (D.15) and doing all ν integrations in Equation (D.17), $J_A(v'_1, \ldots, v'_\nu)$ is of the form

$$J_A(v'_1, \ldots, v'_\nu) = \frac{1}{\sqrt{\kappa_2}^{\nu}} \left(\text{polynomial in } (v'_1, \ldots, v'_\nu)\right) \prod_{k=1}^{\nu} \phi(v'_k), \tag{D.20}$$

where the coefficients of the polynomial depend on γ_1, γ_2, and a only. Using this result in Equation (D.16), the remaining integrals can then be solved using Identities 4, 5, 6, and 7 from Appendix B. The final result is of the form

$$I_A(x) = \sqrt{\kappa_2}^{\|A\|_1} \sum_{i=0}^{\nu} \left(\text{polynomial in } v'_0\right)_i [\phi(v'_0)]^{\nu-i} [1 - \Phi(v'_0)]^i, \tag{D.21}$$

where again the coefficients of the polynomials depend on γ_1, γ_2, and a only.

4.2 Doing the Substitution

The result for $I_A(x)$ obtained thus far does not require any more integrations. However, it does depend on $v'_0 = Q^{-1}(\Phi(x))/\sqrt{\kappa_2+\vartheta^2}$ rather than directly on x. The probability distribution Q of the measured values of the sample is the convolution of P and a normal distribution with mean zero and with variance ϑ^2. Therefore, its cumulants equal the sum of the cumulants κ_k of P and those of the normal distribution. More specifically, distribution Q has cumulants 0, $\kappa_2 + \vartheta^2$, κ_3, κ_4, ... Therefore, the skewness and kurtosis of the distribution are $\kappa_3/\sqrt{\kappa_2+\vartheta^2}^3 = a^3\gamma_1$ and $\kappa_4/(\kappa_2+\vartheta^2)^2 = a^4\gamma_2$, where γ_1 and γ_2 are the skewness and kurtosis of distribution P. Expanding $Q^{-1}(\Phi(x))$ into a Cornish-Fisher series yields according to Equation (A.14)

$$\begin{aligned} v_0' &= \frac{Q^{-1}(\Phi(x))}{\sqrt{\kappa_2 + \vartheta^2}} \\ &= x + \frac{\gamma_1}{6} a^3 \mathrm{He}_2(x) + \frac{\gamma_2}{24} a^4 \mathrm{He}_3(x) + \ldots . \end{aligned}$$

By Taylor expansion around x it follows that

$$1 - \Phi(v_0') = 1 - \Phi(x) - \phi(x) \left[\frac{\gamma_1}{6} a^3 \mathrm{He}_2(x) + \frac{\gamma_2}{24} a^4 \mathrm{He}_3(x) + \ldots \right] + \ldots$$

and that

$$\phi(v_0') = \phi(x) \left[1 - x \left(\frac{\gamma_1}{6} a^3 \mathrm{He}_2(x) + \frac{\gamma_2}{24} a^4 \mathrm{He}_3(x) + \ldots \right) + \ldots \right],$$

where all terms represented by dots contain either cumulants higher than the fourth or products of cumulants higher than the second. Binomially expanding powers of these quantities and as before neglecting terms involving higher-order cumulants it follows that

$$v_0'^{\,k} = x^k + k x^{k-1} \left(\frac{\gamma_1}{6} a^3 \mathrm{He}_2(x) + \frac{\gamma_2}{24} a^4 \mathrm{He}_3(x) + \ldots \right) + \ldots, \tag{D.22}$$

that

$$\begin{aligned} \left[1 - \Phi(v_0')\right]^k = [1 - \Phi(x)]^k - k [1 - \Phi(x)]^{k-1} \phi(x) \\ \left[\frac{\gamma_1}{6} a^3 \mathrm{He}_2(x) + \frac{\gamma_2}{24} a^4 \mathrm{He}_3(x) + \ldots \right] + \ldots, \end{aligned} \tag{D.23}$$

and that

$$[\phi(v_0')]^k = [\phi(x)]^k \left[1 - kx \left(\frac{\gamma_1}{6} a^3 \mathrm{He}_2(x) + \frac{\gamma_2}{24} a^4 \mathrm{He}_3(x) + \ldots \right) + \ldots \right]. \tag{D.24}$$

Inserting these expansions in Equation (D.21) results in

$$\begin{aligned} I_A(x) = \sqrt{\kappa_2}^{\|A\|_1} \\ \sum_{i=0}^{\nu} \left[Z_{i,0}^{(A)}(x) + \frac{\gamma_1}{6} Z_{i,1}^{(A)}(x) + \frac{\gamma_2}{24} Z_{i,2}^{(A)}(x) + \ldots \right] [\phi(x)]^{\nu-i} [1 - \Phi(x)]^i, \end{aligned} \tag{D.25}$$

where the

$$Z_{i,j}^{(A)}(x) = \sum_{k \geq 0} \zeta_{i,j}^{(A)}(k) \mathrm{He}_k(x) \tag{D.26}$$

are polynomials in x. The coefficients $\zeta_{i,j}^{(A)}(k)$ that have been obtained using the *Mathematica* program in Section 6 are listed in Table D.1.

4.3 Putting Things Together

With the simple representation for $I_A(x)$ given by Equation (D.25), we are now in a position to compute the expectations of sums of products of noisy order statistics as required for finding

Table D.1. Coefficients $\zeta_{i,j}^{(A)}(k)$ occurring in Equation (D.26). All coefficients $\zeta_{i,0}^{(A)}(k)$, $\zeta_{i,1}^{(A)}(k)$, and $\zeta_{i,2}^{(A)}(k)$ not listed here are zero. See Equation (D.18) for a definition of the noise coefficient a.

$\zeta_{0,0}^{(1)}(0) = a$	$\zeta_{0,1}^{(1)}(1) = a^2(3 - 2a^2)$	$\zeta_{0,2}^{(1)}(2) = a^3(4 - 3a^2)$
$\zeta_{0,0}^{(11)}(0) = a^2/2$	$\zeta_{0,1}^{(11)}(1) = a^3(3 - 2a^2)$	$\zeta_{0,2}^{(11)}(2) = a^4(4 - 3a^2)$
$\zeta_{0,0}^{(111)}(0) = a^3/6$	$\zeta_{0,1}^{(111)}(1) = a^4(3 - 2a^2)/2$	$\zeta_{0,2}^{(111)}(2) = a^5(4 - 3a^2)/2$
$\zeta_{0,0}^{(1111)}(0) = a^4/24$	$\zeta_{0,1}^{(1111)}(1) = a^5(3 - 2a^2)/6$	$\zeta_{0,2}^{(1111)}(2) = a^6(4 - 3a^2)/6$
$\zeta_{0,0}^{(2)}(1) = a^2$	$\zeta_{0,1}^{(2)}(2) = 2a^3(3 - 2a^2)$	$\zeta_{0,2}^{(2)}(3) = 2a^4(4 - 3a^2)$
$\zeta_{1,0}^{(2)}(0) = 1$	$\zeta_{0,1}^{(2)}(0) = 2a(3 - a^4)$	$\zeta_{0,2}^{(2)}(1) = 6a^2(2 - a^4)$
$\zeta_{0,0}^{(21)}(1) = a^3$	$\zeta_{0,1}^{(21)}(2) = 3a^4(3 - 2a^2)$	$\zeta_{0,2}^{(21)}(3) = 3a^5(4 - 3a^2)$
$\zeta_{1,0}^{(21)}(0) = a$	$\zeta_{0,1}^{(21)}(0) = a^2(6 + 3a^2 - 4a^4)$	$\zeta_{0,2}^{(21)}(1) = 4a^3(3 + 2a^2 - 3a^4)$
	$\zeta_{1,1}^{(21)}(1) = a^2(3 - 2a^2)$	$\zeta_{1,2}^{(21)}(2) = a^3(4 - 3a^2)$
$\zeta_{0,0}^{(211)}(1) = a^4/2$	$\zeta_{0,1}^{(211)}(2) = 2a^5(3 - 2a^2)$	$\zeta_{0,2}^{(211)}(3) = 2a^6(4 - 3a^2)$
$\zeta_{1,0}^{(211)}(0) = a^2/2$	$\zeta_{0,1}^{(211)}(0) = 3a^3(1 + a^2 - a^4)$	$\zeta_{0,2}^{(211)}(1) = a^4(6 + 8a^2 - 9a^4)$
	$\zeta_{1,1}^{(211)}(1) = a^3(3 - 2a^2)$	$\zeta_{1,2}^{(211)}(2) = a^4(4 - 3a^2)$
$\zeta_{0,0}^{(22)}(2) = a^4/2$	$\zeta_{0,1}^{(22)}(3) = 2a^5(3 - 2a^2)$	$\zeta_{0,2}^{(22)}(4) = 2a^6(4 - 3a^2)$
$\zeta_{0,0}^{(22)}(0) = a^4/2$	$\zeta_{0,1}^{(22)}(1) = 2a^3(3 + 6a^2 - 5a^4)$	$\zeta_{0,2}^{(22)}(2) = 12a^4(1 + 2a^2 - 2a^4)$
$\zeta_{1,0}^{(22)}(1) = a^2$	$\zeta_{1,1}^{(22)}(2) = 2a^3(3 - 2a^2)$	$\zeta_{0,2}^{(22)}(0) = 6a^4(2 - a^4)$
$\zeta_{2,0}^{(22)}(0) = 1/2$	$\zeta_{1,1}^{(22)}(0) = 2a(3 - a^4)$	$\zeta_{1,2}^{(22)}(3) = 2a^4(4 - 3a^2)$
		$\zeta_{1,2}^{(22)}(1) = 6a^2(2 - a^4)$
$\zeta_{0,0}^{(3)}(2) = a^3$	$\zeta_{0,1}^{(3)}(3) = 3a^4(3 - 2a^2)$	$\zeta_{0,2}^{(3)}(4) = 3a^5(4 - 3a^2)$
$\zeta_{0,0}^{(3)}(0) = 3a$	$\zeta_{0,1}^{(3)}(1) = 3a^2(9 - 2a^2 - 2a^4)$	$\zeta_{0,2}^{(3)}(2) = 3a^3(16 - 3a^2 - 6a^4)$
	$\zeta_{1,1}^{(3)}(0) = 6$	$\zeta_{0,2}^{(3)}(0) = 6a(4 - a^6)$
$\zeta_{0,0}^{(31)}(2) = a^4$	$\zeta_{0,1}^{(31)}(3) = 4a^5(3 - 2a^2)$	$\zeta_{0,2}^{(31)}(4) = 4a^6(4 - 3a^2)$
$\zeta_{0,0}^{(31)}(0) = 3a^2$	$\zeta_{0,1}^{(31)}(1) = 2a^3(18 - 3a^2 - 5a^4)$	$\zeta_{0,2}^{(31)}(2) = 2a^4(30 - a^2 - 15a^4)$
	$\zeta_{1,1}^{(31)}(0) = 6a$	$\zeta_{0,2}^{(31)}(0) = 4a^2(6 + 2a^4 - 3a^6)$
$\zeta_{0,0}^{(4)}(3) = a^4$	$\zeta_{0,1}^{(4)}(4) = 4a^5(3 - 2a^2)$	$\zeta_{0,2}^{(4)}(5) = 4a^6(4 - 3a^2)$
$\zeta_{0,0}^{(4)}(1) = 6a^2$	$\zeta_{0,1}^{(4)}(2) = 12a^3(6 - 2a^2 - a^4)$	$\zeta_{0,2}^{(4)}(3) = 12a^4(10 - 3a^2 - 3a^4)$
$\zeta_{1,0}^{(4)}(0) = 3$	$\zeta_{0,1}^{(4)}(0) = 12a(5 - a^4)$	$\zeta_{0,2}^{(4)}(1) = 12a^2(14 - 3a^4 - 2a^6)$
		$\zeta_{1,2}^{(4)}(0) = 24$

expected moments of the truncated sample. Inserting Equation (D.25) in Equation (D.13) yields

$$\mathrm{E}[S_A] = \sqrt{\kappa_2}^{\|A\|_1} \frac{\mu!}{(\mu-\nu)!} \frac{\lambda-\mu}{\sqrt{2\pi}} \binom{\lambda}{\mu} \sum_{i=0}^{\nu} \int_{-\infty}^{\infty} \left[Z_{i,0}^{(A)}(x) + \frac{\gamma_1}{6} Z_{i,1}^{(A)}(x) + \frac{\gamma_2}{24} Z_{i,2}^{(A)}(x) + \ldots \right] \mathrm{e}^{-\frac{1}{2}x^2} [\phi(x)]^{\nu-i} [\Phi(x)]^{\lambda-\mu-1} [1-\Phi(x)]^{\mu-\nu+i} \mathrm{d}x.$$

The remaining integration usually cannot be done analytically. Instead, we choose to define coefficients

$$h_{\mu,\lambda}^{i,k} = \frac{\lambda-\mu}{\sqrt{2\pi}} \binom{\lambda}{\mu} \int_{-\infty}^{\infty} \mathrm{He}_k(x) \mathrm{e}^{-\frac{1}{2}x^2} [\phi(x)]^i [\Phi(x)]^{\lambda-\mu-1} [1-\Phi(x)]^{\mu-i} \mathrm{d}x \tag{D.27}$$

that can be computed numerically. Note the close connection to the coefficients $e_{\mu,\lambda}^{i,k}$ defined by Beyer [20] that differ only in that Beyer uses x^k rather than $\mathrm{He}_k(x)$ in the definition, and that can therefore be expressed as linear combinations of the $h_{\mu,\lambda}^{i,k}$ and vice versa. With these coefficients it follows

$$\mathrm{E}[S_A] = \sqrt{\kappa_2}^{\|A\|_1} \frac{\mu!}{(\mu-\nu)!} \sum_{i=0}^{\nu} \sum_{k \geq 0} \left[\zeta_{i,0}^{(A)}(k) + \frac{\gamma_1}{6} \zeta_{i,1}^{(A)}(k) + \frac{\gamma_2}{24} \zeta_{i,2}^{(A)}(k) + \ldots \right] h_{\mu,\lambda}^{\nu-i,k} \tag{D.28}$$

for the expected sum of products of noisy order statistics with exponents given by A. Using Equation (D.28), the expected mean and the expected central moments of the truncated sample can now be determined using the results from Section 1.

The Expected Mean

Taking the expectation of Equation (D.1), it follows directly that the expected mean of the truncated sample is

$$\begin{aligned} \mathrm{E}[m_1] &= \frac{1}{\mu} \mathrm{E}[S_1] \\ &= \kappa_2^{1/2} \left(a_{1,0} + \frac{\gamma_1}{6} a_{1,1} + \frac{\gamma_2}{24} a_{1,2} + \ldots \right), \end{aligned} \tag{D.29}$$

where the coefficients

$$a_{1,j} = \sum_{k \geq 0} \sum_{i=0}^{1} \zeta_{i,j}^{(1)}(k) h_{\mu,\lambda}^{1-i,k}$$

have been obtained using Table D.1 and are listed in Table D.2.

The Expected Variance

Taking the expectation of Equation (D.2), it follows that the expected variance of the truncated sample is

$$\begin{aligned} \mathrm{E}[m_2] &= \frac{\mu-1}{\mu^2} \mathrm{E}[S_2] - \frac{2}{\mu^2} \mathrm{E}[S_{1,1}] \\ &= \kappa_2 \frac{\mu-1}{\mu} \left(a_{2,0} + \frac{\gamma_1}{6} a_{2,1} + \frac{\gamma_2}{24} a_{2,2} + \ldots \right), \end{aligned} \tag{D.30}$$

where the coefficients

$$a_{2,j} = \sum_{k\geq 0} \left(\sum_{i=0}^{1} \varsigma_{i,j}^{(2)}(k) h_{\mu,\lambda}^{1-i,k} - 2 \sum_{i=0}^{2} \varsigma_{i,j}^{(11)}(k) h_{\mu,\lambda}^{2-i,k} \right)$$

have been obtained using Table D.1 and are listed in Table D.2.

The Expected Third Central Moment

Taking the expectation of Equation (D.3), it follows that the expected third central moment of the truncated sample is

$$\begin{aligned} \mathrm{E}[m_3] &= \frac{(\mu-1)(\mu-2)}{\mu^3}\mathrm{E}[S_3] - 3\frac{\mu-2}{\mu^3}\mathrm{E}[S_{2,1}] + \frac{12}{\mu^3}\mathrm{E}[S_{1,1,1}] \\ &= \kappa_2^{3/2}\frac{(\mu-1)(\mu-2)}{\mu^2}\left(a_{3,0} + \frac{\gamma_1}{6}a_{3,1} + \frac{\gamma_2}{24}a_{3,2} + \dots\right), \end{aligned} \qquad \text{(D.31)}$$

where again the coefficients

$$a_{3,j} = \sum_{k\geq 0} \left(\sum_{i=0}^{1} \varsigma_{i,j}^{(3)}(k) h_{\mu,\lambda}^{1-i,k} - 3 \sum_{i=0}^{2} \varsigma_{i,j}^{(21)}(k) h_{\mu,\lambda}^{2-i,k} + 12 \sum_{i=0}^{3} \varsigma_{i,j}^{(111)}(k) h_{\mu,\lambda}^{3-i,k} \right)$$

have been obtained using Table D.1 and are listed in Table D.2.

The Expected Fourth Central Moment and the Expected Square of the Variance

Finally, taking the expectation of Equations (D.4) and (D.5), it follows that

$$\begin{aligned} \mathrm{E}[m_4 - 3m_2^2] &= \frac{\mu^3 - 7\mu^2 + 12\mu - 6}{\mu^4}\mathrm{E}[S_4] - 4\frac{\mu^2 - 6\mu + 6}{\mu^4}\mathrm{E}[S_{3,1}] \\ &\quad - 6\frac{\mu^2 - 4\mu + 6}{\mu^4}\mathrm{E}[S_{2,2}] + 24\frac{\mu-3}{\mu^4}\mathrm{E}[S_{2,1,1}] - \frac{144}{\mu^4}\mathrm{E}[S_{1,1,1,1}] \\ &= \kappa_2^2\left[\frac{(\mu-1)(\mu-2)(\mu-3)}{\mu^3}\left(a_{4a,0} + \frac{\gamma_1}{6}a_{4a,1} + \frac{\gamma_2}{24}a_{4a,2} + \dots\right)\right. \\ &\quad \left. - \frac{\mu-1}{\mu^2}\left(a_{4b,0} + \frac{\gamma_1}{6}a_{4b,1} + \frac{\gamma_2}{24}a_{4b,2} + \dots\right)\right], \end{aligned} \qquad \text{(D.32)}$$

where again the coefficients

$$\begin{aligned} a_{4a,j} = \sum_{k\geq 0} \Bigg(& \sum_{i=0}^{1} \varsigma_{i,j}^{(4)}(k) h_{\mu,\lambda}^{1-i,k} - 4 \sum_{i=0}^{2} \varsigma_{i,j}^{(31)}(k) h_{\mu,\lambda}^{2-i,k} - 6 \sum_{i=0}^{2} \varsigma_{i,j}^{(22)}(k) h_{\mu,\lambda}^{2-i,k} \\ & + 24 \sum_{i=0}^{3} \varsigma_{i,j}^{(211)}(k) h_{\mu,\lambda}^{3-i,k} - 144 \sum_{i=0}^{4} \varsigma_{i,j}^{(1111)}(k) h_{\mu,\lambda}^{4-i,k} \Bigg) \end{aligned}$$

and

$$a_{4b,j} = \sum_{k\geq 0} \left(\sum_{i=0}^{1} \varsigma_{i,j}^{(4)}(k) h_{\mu,\lambda}^{1-i,k} - 4 \sum_{i=0}^{2} \varsigma_{i,j}^{(31)}(k) h_{\mu,\lambda}^{2-i,k} + 6 \sum_{i=0}^{2} \varsigma_{i,j}^{(22)}(k) h_{\mu,\lambda}^{2-i,k} \right)$$

have been obtained using Table D.1 and are listed in Table D.2.

Table D.2. Coefficients occurring in Equations (D.29), (D.30), (D.31), and (D.32). See Equation (D.27) for a definition of the $h^{i,k}_{\mu,\lambda}$ and Equation (D.18) for a definition of the noise coefficient a.

$$a_{1,0} = ah^{1,0}_{\mu,\lambda}$$

$$a_{1,1} = a^2(3 - 2a^2)h^{1,1}_{\mu,\lambda}$$

$$a_{1,2} = a^3(4 - 3a^2)h^{1,2}_{\mu,\lambda}$$

$$a_{2,0} = 1 - a^2(h^{2,0}_{\mu,\lambda} - h^{1,1}_{\mu,\lambda})$$

$$a_{2,1} = 2a(3 - a^4)h^{1,0}_{\mu,\lambda} - 2a^3(3 - 2a^2)(h^{2,1}_{\mu,\lambda} - h^{1,2}_{\mu,\lambda})$$

$$a_{2,2} = 6a^2(2 - a^4)h^{1,1}_{\mu,\lambda} - 2a^4(4 - 3a^2)(h^{2,2}_{\mu,\lambda} - h^{1,3}_{\mu,\lambda})$$

$$a_{3,0} = a^3(h^{1,2}_{\mu,\lambda} - 3h^{2,1}_{\mu,\lambda} + 2h^{3,0}_{\mu,\lambda})$$

$$\begin{aligned} a_{3,1} = {} & 6 - 9a^2(2 - a^2)(h^{2,0}_{\mu,\lambda} - h^{1,1}_{\mu,\lambda}) \\ & +3a^4(3 - 2a^2)(h^{1,3}_{\mu,\lambda} - 3h^{2,2}_{\mu,\lambda} + 2h^{3,1}_{\mu,\lambda} + h^{1,1}_{\mu,\lambda} - 2h^{2,0}_{\mu,\lambda}) \end{aligned}$$

$$\begin{aligned} a_{3,2} = {} & 6a(4 - a^6)h^{1,0}_{\mu,\lambda} + 12a^3(3 - 2a^2)(h^{1,2}_{\mu,\lambda} - h^{2,1}_{\mu,\lambda}) \\ & +3a^5(4 - 3a^2)(h^{1,4}_{\mu,\lambda} - 3h^{2,3}_{\mu,\lambda} + 2h^{3,2}_{\mu,\lambda} - 4h^{2,1}_{\mu,\lambda} + 2h^{1,2}_{\mu,\lambda}) \end{aligned}$$

$$a_{4a,0} = a^4(h^{1,3}_{\mu,\lambda} - 3h^{2,0}_{\mu,\lambda} - 7h^{2,2}_{\mu,\lambda} + 12h^{3,1}_{\mu,\lambda} - 6h^{4,0}_{\mu,\lambda})$$

$$\begin{aligned} a_{4a,1} = {} & 18a^3(2 - a^2)(h^{1,2}_{\mu,\lambda} - 3h^{2,1}_{\mu,\lambda} + 2h^{3,0}_{\mu,\lambda}) \\ & +2a^5(3 - 2a^2)(3h^{1,2}_{\mu,\lambda} + 2h^{1,4}_{\mu,\lambda} - 25h^{2,1}_{\mu,\lambda} - 14h^{2,3}_{\mu,\lambda} \\ & \qquad +18h^{3,0}_{\mu,\lambda} + 24h^{3,2}_{\mu,\lambda} - 12h^{4,1}_{\mu,\lambda}) \end{aligned}$$

$$\begin{aligned} a_{4a,2} = {} & 24 + 48a^2(2 - a^2)(h^{1,1}_{\mu,\lambda} - h^{2,0}_{\mu,\lambda}) \\ & +8a^4(3 - 2a^2)(2h^{1,1}_{\mu,\lambda} + 3h^{1,3}_{\mu,\lambda} - 5h^{2,0}_{\mu,\lambda} - 9h^{2,2}_{\mu,\lambda} + 6h^{3,1}_{\mu,\lambda}) \\ & +4a^6(4 - 3a^2)(2h^{1,1}_{\mu,\lambda} + 3h^{1,3}_{\mu,\lambda} + h^{1,5}_{\mu,\lambda} - 7h^{2,0}_{\mu,\lambda} - 22h^{2,2}_{\mu,\lambda} \\ & \qquad -7h^{2,4}_{\mu,\lambda} + 18h^{3,1}_{\mu,\lambda} + 12h^{3,3}_{\mu,\lambda} - 6h^{4,2}_{\mu,\lambda}) \end{aligned}$$

$$a_{4b,0} = 6 + 12a^2(h^{1,1}_{\mu,\lambda} - h^{2,0}_{\mu,\lambda}) + a^4(h^{1,3}_{\mu,\lambda} + 3h^{2,0}_{\mu,\lambda} - h^{2,2}_{\mu,\lambda})$$

$$\begin{aligned} a_{4b,1} = {} & 72ah^{1,0}_{\mu,\lambda} + 24a^3(h^{2,1}_{\mu,\lambda} - h^{1,2}_{\mu,\lambda} - 2h^{1,0}_{\mu,\lambda}) \\ & +6a^3(2 - a^2)(4h^{1,0}_{\mu,\lambda} + 11h^{1,2}_{\mu,\lambda} - 11h^{2,1}_{\mu,\lambda}) \\ & +2a^5(3 - 2a^2)(3h^{1,2}_{\mu,\lambda} + 2h^{1,4}_{\mu,\lambda} + 5h^{2,1}_{\mu,\lambda} - 2h^{2,3}_{\mu,\lambda}) \end{aligned}$$

$$\begin{aligned} a_{4b,2} = {} & 24 - 24a^2(3h^{1,1}_{\mu,\lambda} + h^{1,3}_{\mu,\lambda} - h^{2,2}_{\mu,\lambda}) \\ & +12a^2(2 - a^2)(13h^{1,1}_{\mu,\lambda} + h^{1,3}_{\mu,\lambda} - 4h^{2,0}_{\mu,\lambda} - h^{2,2}_{\mu,\lambda}) \\ & +4a^4(3 - 2a^2)(13h^{1,1}_{\mu,\lambda} + 15h^{1,3}_{\mu,\lambda} + 2h^{2,0}_{\mu,\lambda} - 15h^{2,2}_{\mu,\lambda}) \\ & +4a^6(4 - 3a^2)(2h^{1,1}_{\mu,\lambda} + 3h^{1,3}_{\mu,\lambda} + h^{1,5}_{\mu,\lambda} - h^{2,0}_{\mu,\lambda} + 2h^{2,2}_{\mu,\lambda} - h^{2,4}_{\mu,\lambda}) \end{aligned}$$

---

5. A Corollary for Normal Samples

Two results required in Chapter 5 now easily follow as corollaries. For samples x_k, $k = 1, \ldots, \lambda$, drawn from a standardized normal distribution with noise of strength ϑ, the expected mean of those μ sample members with the largest associated measured values is

$$\mathrm{E}\left[\frac{1}{\mu}\sum_{i=1}^{\mu} x_{i;\lambda}\right] = \frac{1}{\mu}\mathrm{E}[S_1] = \frac{h_{\mu,\lambda}^{1,0}}{\sqrt{1+\vartheta^2}}, \tag{D.33}$$

where Equation (D.28) and Table D.1 have been used. Note that as the distribution of the x_k is normal, no expansion of the distribution function is required and this result is exact. Likewise, the expected squared mean of those μ sample members with the largest associated measured values is

$$\mathrm{E}\left[\left(\frac{1}{\mu}\sum_{i=1}^{\mu} x_{i;\lambda}\right)^2\right] = \frac{1}{\mu^2}\mathrm{E}[S_2 + 2S_{1,1}] = \frac{1}{\mu}\left(1 + \frac{h_{\mu,\lambda}^{1,1} + (\mu-1)h_{\mu,\lambda}^{2,0}}{1+\vartheta^2}\right), \tag{D.34}$$

where $h_{\mu,\lambda}^{0,0} = 1$ and again Equation (D.28) and Table D.1 have been used.

6. Mathematica Code

This section contains the *Mathematica* code used to compute the coefficients in Table D.1. Let us first define Hermite polynomials:

```
Hermite[k_, x_] := Simplify[HermiteH[k, x/Sqrt[2]]/Sqrt[2]^k];
```

This is necessary as *Mathematica*'s built-in definition `HermiteH` differs from our definition made in Section 4.1 of Appendix A. So as to compose the integrand in Equation (D.17) we define

```
MakePi[A_] :=
  Apply[Plus,
    Map[Apply[Times, MapIndexed[x[First[#2]]^#1 &, #1]] &,
        Permutations[A]]];
```

where `A` stands for the exponent vector A, to compute $\pi_A(x_1, \ldots, x_\nu)$ and subsequently

```
MakeIntegrand1[A_] :=
  MakePi[A]*Product[1+g1(Hermite[3, x[i]])+g2(Hermite[4, x[i]]),
                    {i, 1, Length[A]}];
```

to obtain the integrand. Here, `g1` and `g2` stand for $\gamma_1/6$ and $\gamma_2/24$, respectively. Note that for simplicity we have omitted the terms involving exponential functions from the integrand and that therefore we will need to give integration rules that take that into account.

Before implementing the proper integration routines, let us define a useful routine for expanding polynomials in terms of their Hermite basis:

```
HermiteExpand[expr_, x_] :=
  ToHermite[Expand[expr] /. {g1^i_->0, g1^i_.*g2^j_.->0, g2^i_->0}, x];
```

```
ToHermite[expr1_+expr2_, x_] :=
  ToHermite[expr1, x]+ToHermite[expr2, x];
ToHermite[expr_, x_] :=
  expr*He[0, x] /; Not[MatchQ[expr, a_.*x^k_.]];
ToHermite[expr_. x_^k_., x_] :=
  expr*He[k, x]+ToHermite[Expand[expr*(x^k-Hermite[k, x])], x];
```

HermiteExpand both expresses powers of the x[i] as linear combinations of Hermite polynomials and it eliminates terms involving higher-order cumulants. Note that we use He rather than Hermite as we do not want *Mathematica* to reverse the effect of applying ToHermite. Writing x[i] and y[i] for the u_i' and the v_i', respectively, the ν-fold integration leading to the representation given in Equation (D.20) is then handled by

```
Integrate1[expr_, 0] := expr;
Integrate1[expr_, i_] :=
  Integrate1[Int1[HermiteExpand[expr, x[i]], x[i], y[i]], i-1];

Int1[c_ expr_, x_, y_] := c Int1[expr, x, y] /; FreeQ[c, x];
Int1[expr1_+expr2_, x_, y_] := Int1[expr1, x, y] + Int1[expr2, x, y];
Int1[He[k_, x_], x_, y_] := a^(k+1) Hermite[k, y] g[y];
```

where Int1 implements the integration rule Equation (D.19) taking into account that the terms involving exponential functions had been left out, and where the second argument in the initial call to Integrate1 needs to be ν. Here and in what follows, f[x] and g[x] stand for $1 - \Phi(x)$ and $\phi(x)$, respectively. The integrand in Equation (D.16) is then simply given by:

```
MakeIntegrand2[A_] := Integrate1[MakeIntegrand1[A], Length[A]];
```

where again A stands for the exponent vector A.

The ν-fold integration in Equation (D.16) is done by

```
Integrate2[expr_, 0] := expr;
Integrate2[expr_, i_] :=
  Integrate2[Int2[HermiteExpand[expr, y[i]], y[i], y[i-1]],i-1];

Int2[c_ expr_, x_, y_] := c Int2[expr, x, y] /; FreeQ[c, x];
Int2[expr1_+expr2_, x_, y_] := Int2[expr1, x, y]+Int2[expr2, x, y];

Int2[He[0, x_] g[x_], x_, y_] := f[y];
Int2[He[0, x_] f[x_] g[x_], x_, y_]:= f[y]^2/2;
Int2[He[1, x_] g[x_]^b_., x_, y_] := g[y]^b/b;
Int2[He[1, x_] f[x_] g[x_]^2, x_, y_] :=
  f[y]g[y]^2/2-Int2[He[0, x] g[x]^3, x, y]/2;
Int2[He[k_, x_] g[x_], x_, y_] := Hermite[k-1, y]g[y];
Int2[He[k_, x_] g[x_]^b_., x_, y_] :=
  (Hermite[k-1, y]g[y]^b/b
    - (b-1)(k-1)Int2[He[k-2, x] g[x]^b, x, y]/b) /; k>=2;
Int2[He[k_, x_] f[x_] g[x_], x_, y_] :=
  Hermite[k-1, y]f[y]g[y] - Int2[He[k-1, x] g[x]^2, x, y] /; k>=1;
Int2[He[k_, x_] f[x_] g[x_]^2, x_, y_] :=
  (Hermite[k-1, y]f[y]g[y]^2/2
```

```
  - (k-1)Int2[He[k-2, x] f[x] g[x]^2, x, y]/2
  - Int2[He[k-1, x] g[x]^3, x, y]/2) /; k>=2;
```

where Int2 implements the integration rules given by Identities 4, 5, 6, and 7 from Appendix B and where the second argument of Integrate2 needs to be ν initially.

The result of the steps so far is the representation of $I_A(x)$ given by Equation (D.21). To do the substitution prescribed by Equations (D.22), (D.23), and (D.24) we define:

```
Substitution[c_] := c /; FreeQ[c, y[0]];
Substitution[expr1_+expr2_] :=
  Substitution[expr1] + Substitution[expr2];
Substitution[expr1_*expr2_] :=
  Substitution[expr1] * Substitution[expr2];
Substitution[f[y[0]]^k_.] :=
  f[x]^k - k f[x]^(k-1)g[x](g1 a^3Hermite[2, x]+g2 a^4Hermite[3, x]);
Substitution[g[y[0]]^k_.] :=
  g[x]^k(1-k x(g1 a^3Hermite[2, x]+g2 a^4Hermite[3, x]));
Substitution[y[0]^k_.] :=
  x^k+k x^(k-1)(g1 a^3Hermite[2, x]+g2 a^4Hermite[3, x]);
```

Finally, the representation of $I_A(x)$ given by Equation (D.25) can be obtained by:

```
MakeSum[A_] :=
  HermiteExpand[
    Substitution[Integrate2[MakeIntegrand2[A], Length[A]]]/a^Length[A],
    x];
```

where the division by a^ν reflects the factors $\sqrt{\kappa_2 + \vartheta^2}$ in Equation (D.16) and $1/\sqrt{\kappa_2}$ in Equation (D.20).

So as to print the coefficients in table form appropriate for entering them in Table D.1, let us define:

```
MakeTable[A_] :=
(
  table = Table[0, {i, 3}, {j, 3}, {k, 6}];
  Map[(i := Switch[#, g2*_, 3, g1*_, 2, _, 1];
       j := # /. {f[x]^k_.*_:>k+1, _:>1};
       k := # /. He[k_, x]*_:>k+1;
       v := DeleteCases[#, g2 | g1 | f[x]^k_. | g[x]^k_. | He[k_, x]];
       table = ReplacePart[table, table[[i, j, k]]+v, {i, j, k}]) &,
      MakeSum[A]];
  table // TableForm
);
```

Calling for example MakeTable[{2,1,1}] then arranges the coefficients in a table with $\zeta_{i,j}^{211}(k)$ in row k and column i.

References

[1] M. Abramowitz and I. A. Stegun, *Handbook of Mathematical Functions*, (Dover, New York, 1965).

[2] E. J. Anderson and M. C. Ferris, "A Direct Search Algorithm for Optimization with Noisy Function Evaluations", *SIAM Journal on Optimization*, 11(3), pages 837-857, (2001).

[3] P. J. Angeline, "The Effects of Noise on Self-Adaptive Evolutionary Optimization", in *Proceedings of the Fifth Annual Conference on Evolutionary Programming*, pages 432-439, (1996).

[4] L. Armijo, "Minimization of Functions Having Lipschitz Continuous First Partial Derivatives", *Pacific Journal of Mathematics*, 16(1), pages 1-3, (1966).

[5] B. C. Arnold, N. Balakrishnan, and H. N. Nagaraja, *A First Course in Order Statistics*, (Wiley, New York, 1992).

[6] D. V. Arnold, "Evolution Strategies in Noisy Environments – A Survey of Existing Work", in L. Kallel, B. Naudts, and A. Rogers (eds.), *Theoretical Aspects of Evolutionary Computing*, pages 239-249, (Springer, Heidelberg, 2001).

[7] D. V. Arnold and H.-G. Beyer, "Efficiency and Self-Adaptation of the $(\mu/\mu_I, \lambda)$-ES in a Noisy Environment", in M. Schoenauer et al. (eds.), *Parallel Problem Solving from Nature – PPSN VI*, pages 39-48, (Springer, Heidelberg, 2000).

[8] D. V. Arnold and H.-G. Beyer, "Local Performance of the $(\mu/\mu_I, \lambda)$-ES in a Noisy Environment", in W. Martin and W. M. Spears (eds.), *Foundations of Genetic Algorithms 6*, pages 127-141, (Morgan-Kaufmann, San Francisco, 2001).

[9] D. V. Arnold and H.-G. Beyer, "Investigation of the (μ, λ)-ES in the Presence of Noise", *Proceedings of the 2001 IEEE Congress on Evolutionary Computation*, pages 332-339, (2001).

[10] D. V. Arnold and H.-G. Beyer, "Local Performance of the $(1+1)$-ES in a Noisy Environment", *IEEE Transactions on Evolutionary Computation*, 6(1), pages 30-41, (2002).

[11] D. V. Arnold and H.-G. Beyer, "Performance Analysis of Evolution Strategies with Multi-Recombination in High-Dimensional $\mathbb{R}^N$-Search Spaces Disturbed by Noise", *Theoretical Computer Science*, in press, (2002).

[12] T. Bäck, *Evolutionary Algorithms in Theory and Practice*, (Oxford University Press, New York, 1996).

[13] T. Bäck and U. Hammel, "Evolution Strategies Applied to Perturbed Objective Functions", in Z. Michalewicz et al. (eds.), *Proceedings of the First IEEE Conference on Evolutionary Computation*, pages 40-45, (1994).

[14] T. Bäck, U. Hammel, and H.-P. Schwefel, "Evolutionary Computation: Comments on the History and Current State", *IEEE Transactions on Evolutionary Computation*, 1(1), pages 3-17, (1997).

[15] T. Bäck and H.-P. Schwefel, "An Overview of Evolutionary Algorithms for Parameter Optimization", *Evolutionary Computation*, 1(1), pages 1-24, (1993).

[16] R. R. Barton and J. S. Ivey, "Nelder-Mead Simplex Modifications for Simulation Optimization", *Management Science*, 42(7), pages 954-973, (1996).

[17] M. Bell and M. C. Pike, "Remark on Algorithm 178", *Communications of the ACM*, 9, pages 684-685, (1966).

[18] H.-G. Beyer, "Toward a Theory of Evolution Strategies: Some Asymptotical Results from the $(1 \overset{+}{,} \lambda)$-Theory", *Evolutionary Computation*, 1(2), pages 165-188, (1993).

[19] H.-G. Beyer, "Towards a Theory of 'Evolution Strategies': Progress Rates and Quality Gain for $(1 \overset{+}{,} \lambda)$-Strategies on (Nearly) Arbitrary Fitness Functions", in Y. Davidor, R. Männer, and H.-P. Schwefel (eds.), *Parallel Problem Solving from Nature – PPSN III*, pages 58-67, (Springer, Heidelberg, 1994).

[20] H.-G. Beyer, "Toward a Theory of Evolution Strategies: The (μ, λ)-Theory", *Evolutionary Computation*, 2(4), pages 381-407, (1995).

[21] H.-G. Beyer, "Toward a Theory of Evolution Strategies: On the Benefit of Sex – the $(\mu/\mu, \lambda)$-Theory", *Evolutionary Computation*, 3(1), pages 81-111, (1995).

[22] H.-G. Beyer, "Toward a Theory of Evolution Strategies: Self-Adaptation", *Evolutionary Computation*, 3(3), pages 311-347, (1996).

[23] H.-G. Beyer, "An Alternative Explanation for the Manner in which Genetic Algorithms Operate", *BioSystems*, 41, pages 1-15, (1997).

[24] H.-G. Beyer, "Mutate Large, But Inherit Small! On the Analysis of Rescaled Mutations in $(\tilde{1}, \tilde{\lambda})$-ES with Noisy Fitness Data", in A. E. Eiben et al. (eds.), *Parallel Problem Solving from Nature – PPSN V*, pages 109-118, (Berlin, Springer, 1998).

[25] H.-G. Beyer, "On the Dynamics of EAs without Selection", in W. Banzhaf and C. Reeves (eds.), *Foundations of Genetic Algorithms* 5, pages 5-26, (Morgan Kaufmann, San Mateo, 1999).

[26] H.-G. Beyer, "Evolutionary Algorithms in Noisy Environments: Theoretical Issues and Guidelines for Practice", *Computer Methods in Mechanics and Applied Engineering*, 186, pages 239-267, (2000).

[27] H.-G. Beyer, *The Theory of Evolution Strategies*, (Springer, Heidelberg, 2001).

[28] H.-G. Beyer and D. V. Arnold, "Fitness Noise and Localization Errors of the Optimum in General Quadratic Fitness Models", in W. Banzhaf et al. (eds.), *Proceedings of the 1999 Genetic and Evolutionary Computation Conference*, pages 817-824, (1999).

[29] H.-G. Beyer and D. V. Arnold, "Theory of Evolution Strategies – A Tutorial", in L. Kallel, B. Naudts, and A. Rogers (eds.), *Theoretical Aspects of Evolutionary Computing*, pages 109-133, (Springer, Heidelberg, 2001).

[30] H.-G. Beyer and K. Deb, "On Self-Adaptive Features in Real-Parameter Evolutionary Algorithms", *IEEE Transactions on Evolutionary Computation*, 5(3), pages 250-270, (2001).

[31] D. M. Bortz and C. T. Kelley, "The Simplex Gradient and Noisy Optimization Problems", in J. T. Borggaard et al. (eds.), *Computational Methods in Optimal Design and Control*, pages 77-90, (Birkhäuser, Boston, 1998).

[32] G. E. P. Box, "Evolutionary Operation: A Method for Increasing Industrial Productivity", *Applied Statistics*, 6(2), pages 81-101, (1957).

[33] G. E. P. Box and N. R. Draper, *Empirical Model Building and Response Surfaces*, (Wiley, New York, 1987).

[34] G. E. P. Box and K. B. Wilson, "On the Experimental Attainment of Optimal Conditions", *Journal of the Royal Statistical Society, Series B*, XIII(1), pages 1-45, (1951).

[35] C. Elster and A. Neumaier, "A Grid Algorithm for Bound Constrained Optimization of Noisy Functions", *IMA Journal of Numerical Analysis*, 15, pages 585-608, (1995).

[36] R. A. Fisher and E. A. Cornish, "The Percentile Points of Distributions Having Known Cumulants", *Technometrics*, 2(2), pages 209-225, (1960).

[37] J. M. Fitzpatrick and J. J. Grefenstette, "Genetic Algorithms in Noisy Environments", *Machine Learning*, 3, pages 101-120, (1988).

[38] P. Gilmore and C. T. Kelley, "An Implicit Filtering Algorithm for Optimization of Functions with Many Local Minima", *SIAM Journal on Optimization*, 5, pages 269-285, (1995).

[39] L. Gruenz and H.-G. Beyer, "Some Observations on the Interaction of Recombination and Self-Adaptation in Evolution Strategies", *Proceedings of the 1999 IEEE Congress on Evolutionary Computation*, pages 639-645, (1999).

[40] L. S. Gurin and L. A. Rastrigin, "Convergence of the Random Search Method in the Presence of Noise", *ARC*, 26, pages 1505-1511, (1965).

[41] U. Hammel and T. Bäck, "Evolution Strategies on Noisy Functions. How to Improve Convergence Properties", in Y. Davidor, R. Männer, and H.-P. Schwefel (eds.), *Parallel Problem Solving from Nature – PPSN III*, pages 159-168, (Springer, Heidelberg, 1994).

[42] N. Hansen, *Verallgemeinerte individuelle Schrittweitenregelung in der Evolutionsstrategie*, (Mensch & Buch, Berlin, 1998).

[43] N. Hansen, "Invariance, Self-Adaptation and Correlated Mutations in Evolution Strategies", in M. Schoenauer et al. (eds.), *Parallel Problem Solving from Nature – PPSN VI*, pages 355-364, (Springer, Heidelberg, 2000).

[44] N. Hansen and A. Ostermeier, "Completely Derandomized Self-Adaptation in Evolution Strategies", *Evolutionary Computation*, 9(2), pages 159-195, (2001).

[45] M. Herdy, "Reproductive Isolation as Strategy Parameter in Hierarchically Organized Evolution Strategies", in R. Männer and B. Manderick (eds.), *Parallel Problem Solving from Nature – PPSN II*, pages 207-217, (Elsevier, Amsterdam, 1992).

[46] W. D. Hillis, "Co-Evolving Parasites Improve Simulated Evolution as an Optimization Procedure", in C. G. Langton et al. (eds.), *Artificial Life* II, pages 313-324, (Addison-Wesley, Redwood City, 1992).

[47] R. Hooke and T. A. Jeeves, " 'Direct Search' Solution of Numerical and Statistical Problems", *Journal of the ACM*, 8, pages 212-229, (1961).

[48] D. G. Humphrey and J. R. Wilson, "A Revised Simplex Search Procedure for Stochastic Simulation Response Surface Optimization", *INFORMS Journal on Computing*, 12(4), pages 272-283, (2000).

[49] C. T. Kelley, "Detection and Remediation of Stagnation in the Nelder-Mead Algorithm Using a Sufficient Decrease Condition", *SIAM Journal on Optimization*, 10(1), pages 43-55, (1999).

[50] C. T. Kelley, *Iterative Methods for Optimization*, (SIAM, Philadelphia, 1999).

[51] J. F. Kenney and E. S. Keeping, *Mathematics of Statistics*, Part Two, Second edition, (Van Nostrand, Princeton, 1951).

[52] J. Kiefer and J. Wolfowitz, "Stochastic Estimation of a Regression Function", *Annals of Mathematical Statistics*, 23, pages 462-466, (1952).

[53] R. M. Lewis, V. Torczon, and M. W. Trosset, "Direct Search Methods: Then and Now", *Journal of Computational and Applied Mathematics*, 124, pages 191-207, (2000).

[54] D. G. Luenberger, *Introduction to Dynamic Systems*, (Wiley, New York, 1979).

[55] S. Markon, D. V. Arnold, T. Bäck, T. Beielstein, and H.-G. Beyer, "Thresholding – a Selection Operator for Noisy ES", *Proceedings of the 2001 IEEE Congress on Evolutionary Computation*, pages 465-472, (2001).

[56] K. I. M. McKinnon, "Convergence of the Nelder-Mead Simplex Method to a Nonstationary Point", *SIAM Journal on Optimization*, 9(1), pages 148-158, (1998).

[57] B. L. Miller and D. E. Goldberg, "Genetic Algorithms, Selection Schemes, and the Varying Effects of Noise", *Evolutionary Computation*, 4(2), pages 113-131, (1997).

[58] J. A. Nelder and R. Mead, "A Simplex Method for Function Minimization", *Computer Journal*, 7, pages 308-313, (1965).

[59] V. Nissen and J. Propach, "Optimization with Noisy Function Evaluations", in A. E. Eiben et al. (eds.), *Parallel Problem Solving from Nature – PPSN V*, pages 159-168, (Springer, Heidelberg, 1998).

[60] V. Nissen and J. Propach, "On the Robustness of Population-Based Versus Point-Based Optimization in the Presence of Noise", *IEEE Transactions on Evolutionary Computation*, 2(3), pages 107-119, (1998).

[61] A. I. Oyman, H.-G. Beyer, and H.-P. Schwefel, "Where Elitists Start Limping: Evolution Strategies at Ridge Functions", in A. E. Eiben et al. (eds.), *Parallel Problem Solving from Nature – PPSN V*, pages 34-43, (Springer, Heidelberg, 1998).

[62] M. J. D. Powell, "Direct Search Algorithms for Optimization Calculations", *Acta Numerica*, 7, pages 287-336, (1998).

[63] W. H. Press, S. A. Teukolsky, W. T. Vetterling, and B. P. Flannery, *Numerical Recipes in C: The Art of Scientific Computing*, Second edition, (Cambridge University Press, Cambridge, 1992).

[64] S. Rana, L. D. Whitley, and R. Cogswell, "Searching in the Presence of Noise", in H. M. Voigt et al. (eds.), *Parallel Problem Solving from Nature – PPSN IV*, pages 198-207, (Springer, Heidelberg, 1996).

[65] M. Rattray and J. L. Shapiro, "Noisy Fitness Evaluation in Genetic Algorithms and the Dynamics of Learning", in R. K. Belew and M. D. Vose (eds.), *Foundations of Genetic Algorithms* 4, pages 117-139, (Morgan Kaufmann, San Mateo, 1997).

[66] I. Rechenberg, *Evolutionsstrategie: Optimierung technischer Systeme nach den Prinzipien der biologischen Evolution*, (Frommann-Holzboog, Stuttgart, 1973).

[67] I. Rechenberg, *Evolutionsstrategie '94*, (Frommann-Holzboog, Stuttgart, 1994).

[68] A. Rényi, *Probability Theory*, (North-Holland, Amsterdam, 1970).

[69] H. Robbins and S. Monro, "A Stochastic Approximation Method", *Annals of Mathematical Statistics*, 29, pages 400-407, (1951).

[70] G. Rudolph, "Global Optimization by Means of Distributed Evolution Strategies", in H.-P. Schwefel and R. Männer (eds.), *Parallel Problem Solving from Nature – PPSN I*, pages 209-213, (Springer, Berlin, 1990).

[71] G. Rudolph, "On Correlated Mutations in Evolution Strategies", in R. Männer and B. Manderick (eds.), *Parallel Problem Solving from Nature – PPSN II*, pages 105-114, (Elsevier, Amsterdam, 1992).

[72] G. Rudolph, *Convergence Properties of Evolutionary Algorithms*, (Dr. Kovač, Hamburg, 1997).

[73] G. Rudolph, "Local Convergence Rates of Simple Evolutionary Algorithms with Cauchy Mutations", *IEEE Transactions on Evolutionary Computation*, 1(4), pages 249-258, (1997).

[74] R. Salomon, "Evolutionary Search and Gradient Search: Similarities and Differences", *IEEE Transaction on Evolutionary Computation* 2(2), pages 45-55, (1998).

[75] R. Salomon, "Accelerating the Evolutionary-Gradient-Search Procedure: Individual Step Sizes", in A. E. Eiben et al. (eds.), *Parallel Problem Solving from Nature – PPSN V*, pages 408-417, (Berlin, Springer, 1998).

[76] H.-P. Schwefel, *Numerische Optimierung von Computer-Modellen mittels der Evolutionsstrategie*, (Birkhäuser, Basel, 1977).

[77] H.-P. Schwefel, *Evolution and Optimum Seeking*, (Wiley, New York, 1995).

[78] J. C. Spall, "Multivariate Stochastic Approximation Using a Simultaneous Perturbation Gradient Approximation", *IEEE Transactions on Automatic Control*, 37, pages 332-341, (1992).

[79] J. C. Spall, "Implementation of the Simultaneous Perturbation Algorithm for Stochastic Optimization", *IEEE Transactions on Aerospace and Electronic Systems*, 34(3), pages 817-823, (1998).

[80] J. C. Spall, "An Overview of the Simultaneous Perturbation Method for Efficient Optimization", *Johns Hopkins APL Technical Digest*, 19(4), pages 482-492, (1998).

[81] J. C. Spall, "Adaptive Stochastic Approximation by the Simultaneous Perturbation Method", *IEEE Transactions on Automatic Control*, 45(10), pages 1839-1853, (2000).

[82] J. C. Spall, S. D. Hill, and D. R. Stark, "Theoretical Comparisons of Evolutionary Computation and Other Optimization Approaches", *Proceedings of the 1999 IEEE Congress on Evolutionary Computation*, pages 1398-1405, (1998).

[83] W. Spendley, G. R. Hext, and F. R. Himsworth, "Sequential Application of Simplex Designs in Optimisation and Evolutionary Operation", *Technometrics*, 4, pages 441-461, (1962).

[84] A. Stuart and J. K. Ord, *Kendall's Advanced Theory of Statistics*, Sixth edition, Volume 1: Distribution Theory, (Arnold, London, 1994).

[85] H. Szu and R. Hartley, "Fast Simulated Annealing", *Physics Letters A*, 122, pages 157-162, (1987).

[86] V. J. Torczon, *Multi-Directional Search: A Direct Search Algorithm for Parallel Machines*, Ph.D. thesis, Department of Mathematical Sciences, Rice University, Houston, (1989).

[87] V. Torczon and M. W. Trosset, "From Evolutionary Operation to Parallel Direct Search: Pattern Search Algorithms for Numerical Optimization", *Computing Science and Statistics*, 29, pages 396-401, (1998).

[88] M. W. Trosset, "I Know It When I See It: Toward a Definition of Direct Search Methods", *SIAG/OPT Views and News*, 9, pages 7-10, (1997).

[89] R. E. Walpole and R. H. Myers, *Probability and Statistics for Engineers and Scientists*, Fourth edition, (Macmillan, New York, 1989).

[90] E. W. Weisstein, *CRC Concise Encyclopedia of Mathematics*, (Chapman & Hall, Boca Raton, 1999).

[91] S. Wolfram, *The Mathematica Book*, Fourth edition, (Cambridge University Press, Cambridge, 1999).

[92] M. H. Wright, "Direct Search Methods: Once Scorned, Now Respectable", in D. F. Griffiths and G. A. Watson (eds.), *Numerical Analysis*, pages 191-208, (Addison Wesley, Redwood City, 1995).

Index